D1420409

AT THE WORKS

AT THE WORKS

A STUDY OF A MANUFACTURING TOWN (MIDDLESBROUGH)

by

Florence

LADY BELL

A Reprint with a new Introduction by
Frederick Alderson

DAVID & CHARLES REPRINTS

7153 4350 5

This book was first published by Edward Arnold
in 1907
This edition published 1969
© 1969 new introduction Frederick Alderson

Printed in Great Britain by
Latimer Trend & Company Limited Whitstable
for David & Charles (Publishers) Limited
South Devon House Railway Station
Newton Abbot Devon

INTRODUCTION TO THE 1969 EDITION

My introduction to Lady Bell, DBE, as a writer, came by a somewhat roundabout route. Anticipating a journey to Petra I took up the published letters of that remarkable traveller and scholar Gertrude Lowthian Bell. Among her early letters are a number written from Red Barns, Redcar, and addressed to F.B. or H.B.: they mention ' going to Clarence ' to arrange a nursing lecture, ' returning from Clarence ' and a mothers' meeting, ' the ladies of Clarence ' being friendly and ' unexpected joy; their accounts came right '. Where or what was Clarence, from which, after one visit, she ' came home with papa at 4.35 pm '? Papa, it became clear, was H.B.—for whom Gertrude felt whole-hearted and life-long admiration; F.B. turned out to be his second wife, Florence, who married Hugh Bell, later Sir Hugh, in 1876 when Gertrude was eight. Clarence, in the letters quoted and written when she was just out of her teens, was situated on the north bank of the Tees at a convenient remove from Redcar and, I found, none other than the Bell Brothers ironworks and furnaces at Port Clarence, Middlesbrough.

Gertrude inherited her energy and intelligence from her grandfather, Sir Isaac Lowthian Bell, FRS, builder of the Clarence works and foremost chemical metallurgist of his day. But it was from F.B., her stepmother Lady Bell, that she obviously derived the will and sympathy to maintain, whenever her desert journeys allowed, the round of activities among the ' ladies of Clarence ' or wives of Middlesbrough's ironworkers.

When later I followed these clues to the point of reading Lady Bell's study of that town I discovered that here was a classic description of late nineteenth-century urban life at the centre of a key industry. Middlesbrough provided for study a purely Victorian town. The study was a model, presented with care and modesty, for the fieldwork of future sociologists. Her familiarity with the works side from within and with the workman's lot from an intercourse of nearly thirty years with a large proportion of ironworkers made Lady Bell an ideal observer. Her knowledge of and close family connection with the whole growth of Middlesbrough, from salt-marsh hamlet to scene of ' Titanic industry ' within two generations, gave her an exceptional advantage. In her times the iron trade was regarded as a huge measuring gauge of the national prosperity. She says simply: ' I have tried in this book to put a piece of prosperity under the microscope '. From her own family background— she was the daughter of Sir Joseph Olliffe, MD, physician to the British Embassy at Paris—and from her writing of novels, plays and children's stories, she had a width of understanding. The dedication of AT THE WORKS to Charles Booth, author of *Life and Labour of the People of London*, indicates sympathy and vision also.

When Lowthian Bell and his two brothers, the sons of Thomas Bell, a Newcastle ironmaster, set up their furnaces at Port Clarence (served by the West Hartlepool Railway), they entered the life of Middlesbrough little more than twenty years after that community had itself entered the industrial life of England. Like Scunthorpe, another hamlet swollen by virtue—' though visually it certainly cannot be called virtue ' (Pevsner)—of ironstone and smelting into a composite town of 70,000 ; or Barrow-in-Furness, transformed within half a century of the discovery of its mineral wealth and the development of its railways from a village of 300 souls to an industrial community again of some 70,000, the former Teesside hamlet with its population of only forty in 1829 was to become a large municipal borough by the end of the century. The Bells came in 1853, when Middlesbrough was only one stage on the road to compeerage with such older industrial centres as Rochdale, Burnley, Wigan, whilst Sheffield,

' dark city in a golden frame ', had already a population of 135,000. The twin forces which set the wheels of Middlesbrough's progress turning were coal and Quaker enterprise.

The five Quaker gentlemen who bought the 500 acres of the original Middlesbrough estate, and planned to lay out the site for a town of 5,000 people, did so in order to create a new port for the shipment of coal from South Durham to London and various other markets. New quays were needed, in deeper water and nearer to the sea than the old staiths at Stockton: with an extension of the Stockton & Darlington Railway (opened 1825) Middlesbrough would be on the map. Quaker enterprise proved sound. ' Port Darlington ' was constructed: a small town was laid out, with broad streets and spaces for a modest town hall, market, church, chapel and bank. Within the first ten years coal exports rocketed up from the estimated 10,000 tons annually to 1,500,000 tons. In that time the population rose to 5,463 and progress according to plan included a public market, St Hilda's church and new docks cut into the Tees. A firm of ironfounders, Bolckow and Vaughan, had also decided to open a foundry there.

The pattern of prosperity, however, of this town ' unsupported by the pedestals of time ' was not yet established. Though successful the initial plan was subject to economic subsidence. Within the next ten years Middlesbrough's coal-exporting trade experienced a marked decline as the new network of railways proved able to move coal more conveniently and less expensively than Port Darlington. Middlesbrough might almost have sunk back into the marsh, but—at this point—it entered instead into its second growth stage, one from which its future importance was assured.

The discovery of iron ore at Eston in the Cleveland Hills, or more precisely the recognition of it by Vaughan and his advisers as workable ore, converted Middlesbrough almost overnight from a coal town to an iron town of great national significance. Situated on a navigable river with coalfields within reach and ironstone to hand, no town could have been more favourably placed for an ironmaking centre. Instead of fetching and carrying, as before, Vaughans could now concentrate all their processes close to the source of their raw material.

Once again output soared ahead of estimates. From the 1,000 tons of ironstone expected from Eston each week, when the first blast furnace was blown in 1851, the tonnage quickly reached 3,000 daily: in ten years' time there were over forty furnaces in the district and production of pig-iron amounted to 500,000 tons annually. After another dozen years Middlesbrough's population, swollen by the influx of ironworkers from Durham, Staffordshire, Scotland and South Wales, passed the 40,000 mark, and her puddling furnaces supplied one-third of the whole country's pig-iron. Phenomenal growth indeed; accompanied, as the official handbook admits, by a chaotic rush of inferior house-building, complete streets of mean, badly built terraces (being) sandwiched between the back gardens of the original houses.

It was under this new reign of King Iron that the Bell brothers confidently built up their great enterprise at Port Clarence, and other ironmasters, the Cochranes, Gilkes, Hopkins, Snowdon, Samuelson, staked out their claims. The handbook shows a view of a comparatively primitive works at the beginning of the iron age, and also of the neat and spacious South Street with St Hilda's Church, in 1857, as first planned and built in the ' new town ' between the wharves and the Stockton & Darlington Railway. Compare them for size and complexity with the photograph of a section of Bell Brothers' plant[1]—gantry, bunkers and kilns—and with the town as Lady Bell knew it, with between 800 and 900 streets, the majority of them ' mean ', with over sixty places of worship and twenty-five elementary schools. That is one measure of Middlesbrough's advance,

[1] The early furnaces had a height of 45 to 50 ft, capacity of 5,000 cubic ft and yield of 220 tons weekly ; by the 1870s ' all the small furnaces on the banks of the Tees were demolished ', having been replaced by 80 ft furnaces, capacity 20,000 cu ft and yield of 450 to 550 tons daily. These were standard when Lady Bell wrote.

through its iron age and into the steel age which followed. Eventually the Clarence Works and Associated Collieries were to employ 6,000 people.

As its population and industrial success grew, the grid-iron pattern of streets and municipal buildings that still characterises Middlesbrough spread south, until the new 'new town's' large new Town Hall (opened 1887), shopping area and large new railway station (1877) were no longer in a central position as planned, but themselves almost as much on Middlesbrough's periphery as was the old 'new town' already. And, Lady Bell foresaw, the prosperous citizens of the town who moved their abodes towards the outskirts ' will one day no doubt be engulfed and surrounded by houses'. Old town or new, its brick steadily blackened, not only from the proximity of the furnaces but from the drift from Durham, where the coke ovens alone, in the estimate of Sir Lowthian, sent 45,000 tons of sulphur into the atmosphere each year. 'We are proud of our smoke, an indication of prosperous times,' declared one of the late Victorian mayors.

Between the year 1853, when the original ironfounder Henry Bolckow became mayor, followed four years later by John Vaughan and in 1874 by Thomas Hugh Bell, and that of the latter's retirement from the Council in 1907, Middlesbrough gradually outgrew its ironmasters' proprietorial influence and paternal control. During the iron heyday the town's affairs were largely run by the strong group of founder members of that industry, with their family connections, their weight in council and sense of obligation to their workpeople. But ' it is a matter of ordinary experience that as a man succeeds in his business he moves his home further away from it, and that the distance he lives away from his daily work is in direct proportion to his success in it ' (*At the Works*). The ironmasters were no exception; and as their homes became country halls instead of town houses, their closest links with Middlesbrough's affairs naturally weakened. Industrial changes also had their effect. The self-contained and scientific production of Cleveland pig-iron began to fade after the 1870s, when steel and foreign competition cast its shadow over the ironmasters' world. 'Steel, to the entire exclusion of iron, must be henceforward looked upon as the proper material for railroads, ' asserted Sir Lowthian early in the 80s—when the North Eastern Railway had already ceased buying iron rails. To survive, the ironmasters had to combine or to recapitalise as limited companies with funds drawn from quite different areas and concerns. The new industry was to have a national rather than a local basis; the community, ' perhaps more highly specialised than any which earlier industrial change created ' (Clapham), threw up other leaders.

The steel process invented by Bessemer in 1856 for conversion of non-phosphoric iron had given Barrow, situated up against such an ironfield, a tremendous thrust of growth. For Middlesbrough with its supplies of phosphoric ore the effect was different. Whereas a Haematite Steel Company was in operation near the Walney Channel from 1864 onward and this and other steel centres, with Sheffield in the lead, forged ahead, on Teesside King Iron's dominance remained unbroken. No steel was manufactured there until well into the 70s. When Bolckow-Vaughan, with outside capital and imported material, opened the first Bessemer plant there in 1876 they moved out of a situation with danger signs for Middlesbrough not unlike those shown by the earlier recession of the seaborne coal trade.

Once again the situation changed in Middlesbrough's favour. A process of steel-making from phosphoric iron ore, invented by Gilchrist Thomas, was successfully demonstrated at Bolckow-Vaughan's works (1879). Would-be licensees called on Thomas even before breakfast! Teesside was given a complete new lease of industrial life. Operations started at once, though faster abroad than in Britain —where an organisation now existed for bringing foreign ore from mines close to the sea in Spain to furnaces close to the sea at Middlesbrough and other places. Even so six converters adapted to the new process were in production at the Bolckow-Vaughan plant within three years, and within six years steel tonnage for British shipping had risen almost fivefold. 'No one questions the superiority of steel over iron for naval architecture, ' admitted Lowthian Bell: ' the only barrier to its exclusive use is one of price '. (Bell's authoritative *Principles of the Manufacture of Iron and Steel* appeared in 1884. He received the first Bessemer gold medal.)

While the railway companies abandoned the use of iron, shipbuilders kept up a high demand for iron plates and angles even when nearly all boilers were being made of steel; but by the time *At The Works* was being written the old pattern of the iron town was altered considerably. Dorman and Long had come in, prepared to take advantage of the rising tide of steel and (to look ahead) by a series of mergers and amalgamations, which embraced Bolckow-Vaughan, were destined to become the largest suppliers of structural steel in the kingdom with a worldwide reputation in engineering and bridge building. Already before the century's end they were associated in a limited company with Bell Brothers, who remained mainly iron producers, and Sir Lowthian became chairman of their board. The vast scale of steel plant needed this revised scale of business organisation. In the town also the scale of the ironmasters' original settlement was being lost to sight. Population, recruited by the incessant influx of fresh workers, rose from 40,000 to 90,000 in the last thirty years of the century. As land development spread ever farther south and the ' rows and rows of little brown streets, hastily erected, instantly

occupied ' sprang up, such an old village as Linthorpe
became an indistinguishable part of Middlesbrough
itself.   (The lower end of Linthorpe Road is today a
slum.)   Six years after Lady Bell's survey came out
the Transporter Bridge was built, to link north of the
Tees with south—' the largest bridge of its kind in
the world, 850 feet long, its car can take 600 people '
—replacing the ' cumbrous steam ferry, plying every
quarter of an hour from 5.15 a.m. to 11.0 p.m. between the two sides of the river, filled almost to
overflowing at the time the " shifts " change '.

The ' shifts ', their jobs, their families, their
leisure and their living standards, their drinking
habits, especially the lot of their wives and daughters,
provide the subject of the book for which the above
gives some slight background.   Such a spotlight on
Middlesbrough, which was and is ' both economically and temperamentally closer to the North East
as a whole than to its parent county ', as a recent
writer on Yorkshire remarks, illuminates as a matter
of course working-class conditions in other centres of
Victorian   heavy   industry.   The   peculiarity   of
Middlesbrough is that it was virtually without a past.
Its Victorian development of streets and amenities
had no leaven of ' Georgian architecture, the last
product of a settled and agricultural civilisation when
craftsmanship was understood and enjoyed and
rules of proportion were widely known among buil-
ders ' (Betjeman).   In towns of an older tradition

and history, the newer portion attaches to a more ancient nucleus ' of more dignified dwellings in which a succession of people with a sure foothold in life have [added] to existence not only the useful and essential but that which is able to beautify '. (*At the Works*). Middlesbrough, on the other hand, was a raw town scarcely seventy years old when Lady Bell wrote, with nothing to appeal to a sense of art or beauty— ' no picture gallery, indeed not a picture anywhere that the ordinary public could go to see ', and with ' the more educated in the community, those who should be able to bring a leaven of art, of literature, of thought, to the toilers round them, toiling also themselves '. Like Scunthorpe, of ' grim, grimy centre and High Street with not a single redeeming feature ' (Pevsner), its whole *raison d'etre* and preoccupation was making iron.

What were the products of this relentless environment? What was the health and moral state of this Victorian town that has come, after another sixty years, to be regarded as the commercial and cultural centre of the county borough of Teesside? Was the ironworker's reaction like that of the steel worker, in a ' dreary town ' near the east coast during the 30s where he could earn, so rumour said, £20 a week at the blast furnaces: ' I'm half dead by the weekend. I don't feel like doing anything . . . work and bed, work and bed, week in and week out . . . the pay's good, but what's the use of money, if there's nothing to spend it on? ' (*How the Other Man Lives*, Walter Greenwood). Were the problems of an industrial area at the beginning of the century different or not from those of today, when the call is still ' No room to expand? Short of labour? Then take a look at ESTON for industrial sites—with steel, chemicals, oil, shipbuilding and an expanding deep water dock area'?

For those who wish to take a look, not perhaps at Eston but at the site of old Middlesbrough and its works, a ' perambulation ' can be made which will afford, to the enthusiast in industrial archaeology, as many discoveries and conjectures as the desert traveller is likely to obtain from the lost city of Petra.

Approach Middlesbrough via the Stockton Road, A67, which merges into Newport Road with its congeries of ' mean streets ' adjoining the Ironmasters' District: their names ring back the past—Disraeli St, Gladstone St, Bolckow St, Vaughan St, Gilkes St, Samuelson St and Marsh Road. (Hopkins and Snowdon, the other early ironmasters, are recalled by streets near Depot Road.) On reaching Corporation Road and the new Town Hall—in the grandiose Gothic style favoured by Alfred Waterhouse—turn down Albert Road, cross the railway to Cleveland St (where Henry Bolckow lived in his early days before moving out to a country house), and what remains of the old town lies ahead. Lower South St still carries its original name sign ; from it the blackened spire of St Hilda's and the neat clock tower of the

old, unpretentious Town Hall are still in view.
From the central square of the old town, East St,
West St, South St and North St angled off ; their
symmetrical line is clear and some of the old side-
street names—Henry St (after Bolckow?) for example
—can be found. For the rest, St Hilda's parish
is a redevelopment area, with all that that means.
The church is derelict, the old Town Hall has
acquired a brick block of clinic, library and police
station: matchbox-style three-storey flats occupy the
cleared ground but fail to satisfy any sense of
' township '. Only one thing is unchanged : the
metalliferous air and its sulphurous smell.

On the industrial side enough remains for the
seeing eye to bring former scenes back to life. From
North St proceed past the old Customs House to the
aptly named Vulcan St, busy within living memory :
a goods line, the cinder ballast overgrown with
thistles and willowherb, runs down its length towards
North Wharf. At the lower end the great iron-
banded chimney, tall black buildings, high walls
patterned with ovals of brick and great wooden
gates of the Bolckow-Vaughan works stand, smoke-
less, windowless and empty, but seemingly durable
as iron itself. The office block at the Dock St end
carries now Dorman Long's brass plate. Close by
is the Captain Cook Inn and Ferry Road; from here,
instead of that ' cumbrous steam ferry ' a spectacular
—and equally cumbrous—transporter bridge con-

veys bus, car and pedestrian across the Tees. One can ascend by a zig-zag iron stairway in the super-structure, slippery in the wet, to the girder platform at the top and from there command an extensive view over Teesside.

The tickets issued for this privilege state a three-penny fare from Middlesbrough to Port Clarence. Although the old port on the north bank is gone, parts of High Clarence help out the reconstructive imagination. On the south bank the old wooden staiths are there, once used by Bell Brothers to take Cleveland ironstone by barge to their Clarence Wharf. (After looking at some of the old staiths in the wind-ing, narrow river at Stockton it is clear enough why a new port for coal shipment was needed, and why this ' Middlesbrough Reach ' of the Tees attracted expanding industry in the age of mid-Victorian enterprise.) The prospect, however, from the bridge platform through the ' natural haze '—as a local paper termed it—over the cindery wastes, the gaunt ranks of cranes and black chimneys below the wharfs will be pleasing only to the specialist's eye. In the newer developments further down river, where the modern plants are, today's motto, ' steel is our business ', is spelt out boldly; yesterday's motto, ' the dignity of power ', was expressed more obscure-ly in Middlesbrough's iron beginnings, to judge by what remains.

FREDERICK ALDERSON

AT THE WORKS

A STUDY OF A MANUFACTURING TOWN

AT THE WORKS

A STUDY OF A MANUFACTURING TOWN

BY

LADY BELL

(MRS. HUGH BELL)

AUTHOR OF

'THE MINOR MORALIST,' 'FAIRY-TALE PLAYS,' 'THE ARBITER,'
'WORDLESS CONVERSATION,' ETC.

LONDON
EDWARD ARNOLD

41 & 43 MADDOX STREET, BOND STREET, W.

1907

TO

CHARLES BOOTH

OF WISE AND SYMPATHETIC COUNSEL

The sincere thanks of the writer are due to all those who have helped her to collect the information reproduced in this book.

The chapter on 'Reading' is reprinted, with modifications, from the *Independent Review*, by the kind permission of the Editor.

The figures on pp. 10 and 132 are quoted, also by kind permission, from the *North-Eastern Daily Gazette*.

The Illustrations are reproduced from photographs taken by Mr. W. L. Johnson.

INTRODUCTORY

THE following notes are the outcome of an intercourse of nearly thirty years with a large population
of ironworkers in the north of Yorkshire. During
this time more than a thousand working-men's
homes have been visited, many of them on terms
of friendly and continuous intercourse, by several
female visitors, of whom the writer is one. Wherever
the word 'visitor' is used in the following pages it
refers to one of this small group of eye-witnesses.
The facts recorded can be vouched for. The impressions derived from them may often be mistaken,
although experience should, and usually does in the
long run, enable the visitors to focus and readjust
their impressions with tolerable accuracy.

I have not attempted to deal with the larger
issues connected with the subject, with the great
questions involved in the relations between capital
and labour, employers and employed; I have tried
but to describe, so far as it is possible for an onlooker to do so, the daily lives of the workmen
engaged in carrying on the Iron Trade of this
country in one of its centres of greatest activity.

These workmen are a most important section of the community, not only from their numbers, but from the value of the industry which they represent. The conditions under which they live and work, their consequent possibilities of individual development, mental, moral, and physical, the way their children are brought up, are of vital moment to the country. Such conditions, in these days when so many people try honestly to ascertain what the lives of those in less favoured circumstances than themselves are like, will probably be familiar to a number of those who read this book. In this time of readjustment, the attention of the general public has been more directed, probably, to economic and industrial problems, to varying social conditions, than ever before: and even those unversed in such questions are called upon to discuss and consider them. It may be of interest to those who have never come in contact with a manufacturing population to see what the Iron Trade, which they know but by name, perhaps, as a huge measuring-gauge of the national prosperity, is in reality, when translated into terms of human beings.

We move through a world filled with labels, and we are most of us content to accept the mere name on the label for that which it represents, without seeking to know anything further; we are very glib with certain phrases, which we should be puzzled to define, if we were suddenly, according to Galton's method, asked to 'visualize' them without an instant's pause for thought. Commerce, Prosperity, Industry, the Iron Trade, War, Peace—what do all these mean? I confess that as I try to grasp them I can represent them to myself, always and ever, in terms only of human beings: they all mean the lives, the daily actions, of thousands of our fellow-creatures.

This is the only medium in which we can actually see, handle, realize the great generalizations. We cannot visualize the Nation, let us say, in any other way. It may, indeed, mean something definite to our minds, some great spirit of conduct, a great idea held in common with millions of other people, a name which stands for a purpose, for an object, for a reason, for a nucleus: but we cannot come into personal contact with that idea, although we may think of it and talk about it.

The trade of the nation—what is that? It is only a word, a name. The real trade is, and must be, the trade of individuals.

We read nowadays in every paper discussions as to the best methods of increasing this or that one of our national industries, and in another part of the same paper we may find columns of equally earnest, anxious discussions respecting the deterioration of our race. We are apt to believe that deterioration is likely to be lessened as prosperity increases. But experience does not always bear out this theory.

It may be open to question whether what is called Prosperity is more likely to prevent deterioration or actually to promote it ; whether the very conditions, indeed, arising from 'good times' in commerce, the increasing number of workers who rush to a given place and struggle to live within a given area, do not inevitably make for deterioration at the same time. The conditions of prosperity are no more necessarily beneficial or agreeable to the people whose work is promoting it than the long sea-voyage prescribed for the rest and enjoyment of the man who is well-to-do is necessarily beneficial and agreeable to the stoker who is toiling and panting below.

When we are told that in a given year the Iron Trade was by so many millions of tons more success-ful than the year before, and that so many more thousands of people thereby found a means of liveli-hood, we rejoice, and rightly, over the commercial activity of the country. But while we rejoice, it is well to inquire how many of these thousands are actually made happy by such prosperity : how many of the increasing number of those through the work of whose hands this great output is effected are learning in practice to know the splendid oppor-tunities that life has to offer, not only for the body but for the soul.

I have tried in this book to put a piece of pros-perity under the microscope. That which is revealed in any field of vision by the microscope is obviously that which is really there, although when using our ordinary defective methods of observation we may not see it. But it is there all the same. I have tried to consider, not in general the lot of thousands, but in detail the lives of some of the individuals who compose those thousands; for it is detail that is really convincing, that brings the vivid flash of realization and misgiving. We are gripped and sickened with horror if we see one man killed at our feet by an accident: but we can read with a calm and pitying ejaculation the statistics of the hundreds who have been killed in the same way during the past twelvemonth. If we see in a cottage one emaciated little child wasting away because it has not enough to eat, the sight will make more impression on us than many lists of infant mortality.

This happily does not apply to the sight of misfortune only. A visit to a working man's home of the best kind, where the man is keen about both his work and his leisure, where the woman is a competent manager and a good wife and mother, is more cheering as to the present outlook of the working classes, more inspiring as to their future possibilities, than listening to many orations on the subject, or reading many pages of print.

It is important when endeavouring to describe the life of the ironworkers to dwell upon homes of this kind, as well as on the less fortunate. The latter are naturally brought prominently before our notice, both by the suffering inmates themselves, and by those who would bring them alleviation. It is not so necessary to call attention to those who are prospering. The result is that the needy and unhappy homes appear to preponderate to a greater extent than they do in reality over the happy homes that have no history. This should be borne in mind in reading this book.

CONTENTS

LIST OF ILLUSTRATIONS

AT THE WORKS

CHAPTER I

THE GENESIS OF THE TOWN

THE important part played by the manufacture of iron, and of its further stage of steel, in the industry of a country will be realized when we remember that it forms the basis of all machinery, of all weapons, of all implements used in agriculture and in every craft, of most utensils used in daily life, of all locomotion by sea and by road. It will be seen what an enormously important part all this plays, ever increasingly, in our national life. We are not here, however, considering these various branches, which have been referred to only to show the immense amount of iron required by the country. We are here concerned with the ironworks, the first stage of manufacture, where the iron is extracted from the ironstone brought from the mines. When iron is found in a district, it means that there will be employment not only for the man skilful and deft with his hands, who has a turn, perhaps, for mechanics, for science, for what may be called the higher branches of ironmaking, but there will be employment for countless numbers besides; any man with a pair of strong arms can join in the rush for the new great opportunity. For the majority of the iron-workers, the main equipment needed is health and strength. They must be hale enough and strong enough to lift bars of iron and carry them from one place in the works to another, or to wheel a barrow full of ironstone from the kiln to the furnace.

It is obvious, then, what a field for labour is suddenly opened by the discovery of iron in any part of the country. The genesis of an ironmaking town which follows such a discovery is breathless and tumultuous, and the onslaught of industry which attends the discovery of mineral wealth, whether ironstone or coal mines, has certain characteristics unlike any other form of commercial enterprise. The unexpectedness of it, the change in the condition of the district, which suddenly becomes swamped under a great rush from all parts of the country of people often of the roughest kind, who are going to swell the ranks of unskilled labour; the need for housing these people; all this means that there springs, and too rapidly, into existence a community of a pre-ordained inevitable kind, the

members of which must live near their work. They
must therefore have houses built as quickly as possible; the houses must be cheap, must be as big as
the workman wants, and no bigger; and as they
are built, there arise, hastily erected, instantly
occupied, the rows and rows of little brown streets,
of the aspect that in some place or another is
familiar to all of us. A town arising in this way
cannot wait to consider anything else than time
and space: and none of either must be wasted on
what is merely agreeable to the eye, or even on
what is merely sanitary. There can be no question under these conditions of building model
cottages, or of laying out a district into ideal
settlements. As one owner after another starts
ironworks in the growing place, there is a fresh
inrush of workmen, and day by day the little houses
spring hurriedly into existence, until at last we find
ourselves in the middle of a town. It is, unhappily,
for the most part a side issue for the workman
whether he and his family are going to live under
healthy conditions. The main object of his life is
to be at work: that is the one absolute necessity.

Most of the houses run up in this way consist of
four rooms: two rooms on the ground-floor, one of
them a kitchen and living-room, which in many of
them opens straight from the street, and in some
has a tiny lobby with another door inside it—and
another room behind, sometimes used as a bedroom,

sometimes shut up as a parlour. A little steep dark staircase goes up from the kitchen to the next floor, where there are two more rooms. Sometimes there is a little scullery besides, sometimes a place hardly big enough to be called a room, just big enough to contain a bed, off the kitchen. Such abodes are big enough to house comfortably a couple and two or three children, but not to house the families of ten, twelve, and more, that are sometimes found in them.

Nearly every town of any considerable size that has not remained stagnant has of course in the newer portions of it quarters such as we have described, which, to keep pace with the development of the town, add themselves on to the older part of it. And whether it is a manufacturing quarter added to a town which had none before, or whether it is a whole new manufacturing town where before there was nothing, the conditions of life in such a quarter must be the same until we have found some more desirable way of dealing with them. Such a place must always necessarily present the same general characteristics. The workmen all struggle to be as near as possible to their work, to waste no time or money in transit; they are bound, therefore, to be crowded together, to have no open spaces round them in the places where the majority of them live. There will always be the rows of small cheap houses, the successive streets built as the district prospers, with no point of view but that of affording further dwellings as soon as possible for the throng of workers.

As we look down the unending vista of street after street, all alike, we are somehow apt to consider the aspect as being more discouraging than any one of the houses would look if it stood by itself. But perhaps in this we are mistaken. There is no reason in any sort of street why the life of each individual should be the more monotonous because his next-door neighbour has a front door resembling his own. Life, happily, is not governed only, if at all, by the outward aspect of the home. The dwellers in South Kensington squares and streets who have houses all alike with columned porticoes, may have lives entirely and interestingly differentiated one from another, and so may the dwellers in the small streets of the ironmaking town, where you may open two doors side by side into houses of identically the same accommodation, and find one of them a bright and seemly home and the other an abode hardly fit to be entered by human beings.

There is nothing more difficult, in looking at some one else's house or way of living, than to ascertain exactly what the qualities and defects are from the point of view of the occupant, although it may be easy to see what they are from the point of view of the spectator.

In towns of an older tradition and history, in which the newer portion has been added to a more ancient nucleus, such little monotonous streets as we have described are often erected within measurable distance of something entirely different, of more dignified dwellings in which a succession of people with a secure foothold in life have handed on one to the other a worthy tradition, adding to existence not only the useful and essential, but that which is able to beautify. In such a case the older town insensibly leavens the newer, while it is leavened by it; the older order can help to organize the growth and development of the rough teeming life which lives side by side with it. But where there is no older nucleus of habitation, when the whole community springs into existence at once from the top to the bottom, then, indeed, it is bound to be rough-hewn. The more educated in such a community,—those who should be able to bring a leaven of art, of literature, of thought, to the toilers round them,—are toiling also themselves; they are part of the immense machine, and it is impossible for them to stand aside from it and to judge of it with a free mind. This is the condition of Middlesbrough, the town we are describing, in which leisure, that all-important factor in the development of the mind and soul, is almost unknown: a typical town in which to study the lives of those engaged in the making of iron, for it has come into existence for that purpose and for nothing else.

It is obviously not a place that people would be likely to settle in unless there were very practical reasons for their doing so. There are no immediate surroundings, either of buildings or of country, to appeal to the æsthetic side of imagination, although five or six miles south of the town the beautiful Yorkshire moors begin. There is nothing to appeal to a sense of art or of beauty. There is no building in the town more than seventy years old; most of them, indeed, are barely half that age. There is no picture-gallery: indeed, there is not a picture anywhere that the ordinary public can go to see. But yet imagination can be stirred,—must be stirred,—by the story of the sudden rise of the place, by the Titanic industry with which it deals, by the hardy, strenuous life of the North, the seething vitality of enterprise with which the town began. And to-day all this has consolidated into something more potent still, for the energy of those first fashioners has been transmitted to their successors, and in every layer of the social scale the second generation is now continuing the work of the first. The employers, the professional men, the tradesmen, are handing on their work to the sons who come after them. In many cases the sons of the workmen are going on working for the same employers as their fathers did before them. All this is gradually creating a precedent, a tradition, a spirit of cohesion and of solidarity. Many of the inhabitants of to-day have been born in the town or the district, and to a large proportion of them the associations of home and childhood, of leisure and enjoyment, as well as of work, are already centred in a place which to the first restless generation of new-comers who started it was nothing but a centre of industrial activity.

The whole town, as a place of residence, is designed for the working hours of the people who live in it, and not for their leisure; so far, that is, as the more prosperous inhabitants are concerned. It is a matter of ordinary experience that as a man succeeds in his business he moves his home farther away from it, and that the distance he lives from his daily work is in direct proportion to his success in it. In compliance with this instinctive custom, the more successful and prosperous citizens of the town we are describing have gradually moved their abodes towards the outskirts, and there have grown up round the town detached villas with gardens, forming agreeable semi-country abodes which will one day no doubt be engulfed and surrounded by houses. There are fine public buildings of the kind that grow up with the development of a municipality: a Town Hall, a Free Library, the various offices of the Corporation, churches, schools, a fine park on the outskirts of the town, a big public square in the middle of it, several wide streets with good shops, and towards the south of the town broad roads planted with trees. But towards the centre and the north, serried together in and out of the better quarters, there are hundreds of the little streets we have described, in which lives a struggling, striving, crowded population of workmen and their families, some of them, as will be shown, prospering, anchored, tolerably secure, some in poverty and want, the great majority on the borderland between the two; it is a population recruited by the incessant influx of fresh workers into the town, of which a great part is for ever changing and shifting, restlessly moving from one house to another, or going away altogether, in the constant hope that the mere fact of change must be an improvement. There are altogether in the town between 800 and 900 inhabited streets—including in this term mews and yards teeming with people, as well as streets of various grades and those which are called roads and terraces—and of these nearly 500 are little streets of the kind that a recent writer of fiction has called 'mean streets,' a term that serves well enough for purposes of definition. The working people greatly outnumber the rest of the inhabitants, since they amount to five-sixths of the whole population; and of these one-third are children. But of whatever age or class, all are directly or indirectly concerned in the iron trade. The majority of the population, consisting of the workmen employed at the various works and their employers, are, of course, actively taking part in it: and the rest—the clergymen of various denominations, the doctors, the teachers, the tradesmen—represent all the callings that so large and flourishing a community demands for its spiritual, intellectual, and material needs.

There is in the town, which has already made itself a name for its educational enterprise, an increasing number of large elementary schools, of which there are at present twenty-five.

There are sixty places of worship of one kind and another, of which nine churches and about half a dozen mission-rooms belong to the Church of England; there are three Roman Catholic churches, of which one is the cathedral; twenty Nonconformist places of worship, of which two are Congregational churches, six Baptist, one of the Society of Friends, two Christian Missions, three Presbyterian, three Methodist New Connection, two Plymouth Brethren, five Primitive Methodist, two United Methodist Free Church, nine Wesleyan, one Unitarian; there are also two Salvation Army Missions, one Church Army Mission, and one Spiritualist Association.

It may be of interest to add here that at a moment when the total population was nearly 97,000, the number of people attending any place of worship was 22,190, divided as follows :

Men	...	...	... 7,234
Women	...	...	... 8,360
Children	...	...	... 6,598

That is, there are about 70,000 who attend no place of worship at all.

It is somewhat surprising, when one sees the thronged overflowing Middlesbrough of to-day, to realize that so late as 1801 it only had twenty-five inhabitants. It was no larger, in fact, than the little hamlet of the same name which six or seven centuries before had stood on the banks of the Tees at a place where the monks going between Whitby and Durham were ferried across.

In 1811 the population of Middlesbrough was 35.
„ 1821 „   „   „   „   „ 40.
„ 1831 „   „   „   „   „ 154.
„ 1841 (after railways had begun)   5,463.

In 1850 the ironstone was discovered in the Cleveland Hills, ten miles from Middlesbrough, and then the great town we have described rose on the banks of the Tees, within reach, on the one hand, of the Durham coal-fields, and on the other of the Cleveland ironstone. There could not have been a more favourable place for an ironmaking centre.

The population in 1861, after the discovery of the iron, had risen to 18,892.
The population in 1871 was 39,284
„   „   1881 „ 55,288
„   „   1891 „ 75,532
„   „   1901 „ 91,302

And at the moment of writing the Municipal Borough is over 100,000, and the Parliamentary Borough 116,546.

In default of a romantic past, of a stately tradition, the fact of this swift gigantic, growth has given to the town a romance and dignity of another kind, the dignity of power, of being able to stand erect by its sheer strength on no historic foundation, unsupported by the pedestals of time. And although it may not have the charm and beauty of antiquity, no manufacturing town on the banks of a great river can fail to have an interest and picturesqueness all its own. On either shore rise tall chimneys, great uncouth shapes of kilns and furnaces that appear through the smoke on a winter afternoon like turrets and pinnacles. It might almost be the approach to Antwerp, save that the gloom is constantly pierced by jets of flame from one summit or another, that flare up through the mist and subside again. Twilight and night are the conditions under which to see an ironmaking town, the pillars of cloud by day, the pillars of fire by night; and the way to approach such a town is by the river. No need to inquire into the origin of this eternal highway or its antiquity when we know that it is and must be the oldest of all, and that from time immemorial, before any dwellings at all rose on the banks, it must have been up the river that new-comers from afar held their way.

The Tees, here the boundary between Durham

and Yorkshire, has been widened from two miles above Middlesbrough to the sea, allowing the passage of the shipping which takes the iron to all parts of the world.

There are many different ironworks in the district described, in all of which the life and conditions of the people employed are presumably much the same; the description of one may therefore be taken as applying to all.

Some of these works lie on the north bank of the river. A cumbrous steam ferry, a huge flat boat like an unwieldy floating platform, plies every quarter of an hour, from 5.15 a.m. until 11 p.m., between the two sides of the river, filled almost to overflowing with workmen at the time the ' shifts ' change and the tide sets in of men coming and going to their work. Now and again this ferry is checked in its passage across the river by some stately ship coming in to be loaded with iron at the Middlesbrough wharf, or passing slowly out with its burden. In times of prosperity the river re-echoes with the noise of hammering and machinery from the ship-yards, foundries, and ironworks along the banks. The banks of the Tees are not here, except to the eye of the economist, the ' lovelier Tees ' so dear to Macaulay's exiled Jacobite ; and it is hard to believe that forty miles further up the junction of the Greta with the Tees is still as beautiful, as untouched, as unspoilt, as it was when it was sung by Scott.

But at Middlesbrough we have left verdure far

behind. The great river has here put on its grimy
working clothes, and the banks on either side are
clad in black and grey. Their aspect from the deck
of the ferry-boat is stern, mysterious, forbidding:
hoardings, poles, chimneys, scaffoldings, cranes,
dredging-machines, sheds. The north shore, the
Durham side, is even more desolate than the other,
since it has left the town behind, and the furnaces
and chimneys of the works are interspersed with
great black wastes, black roads, gaunt wooden
palings, blocks of cottages, railway-lines crossing the
roads and suggesting the ever-present danger, and
the ever-necessary vigilance required in the walk
from the boat. A dusty, wild, wide space on which
the road abuts, flanked by the row of the great
furnaces, a space in which engines are going to and
fro, more lines to cross, more dangers to avoid;
a wind-swept expanse, near to which lie a few
straggling rows of cottages.

A colony of workmen live here, actually in the
middle of the works. Westwards from the ferry
there is another settlement, of which the inhabitants,
therefore, are nearer to their work than if they were
on the south side of the river, but not so near as if
they were actually in the heart of the works. Some
more live at another little village about a mile and a
quarter further west from the ferry, and the rest,
the greater number, live on the south side of the
river, in the town.

The river makes a bend at the ferry landing, curving down to the south for about half a mile, and then coming up to the north again. The streets come straggling down towards the river, some of them end-wise towards it, and others where it bends up northwards again parallel to it and facing it. The people living in the latter, in houses so near to the bank that the children playing outside are perilously near the edge, have at any rate the wide expanse of water to look across, and the pure, if keen-edged, wind that blows from it. But the outlook on the other side towards the land is either on to the backs of the little houses opposite, and their yards, or, to those who live at the end of a row, the black plain with the furnaces, trucks, sheds, and scaffolding: houses in which every room is penetrated by the noise of machinery, by the irregular clicking together of trucks coming and going, and by the odours and vapours, more or less endurable according to the different directions of the wind, from the works and coke-ovens. It is a place in which every sense is violently assailed all day long by some manifestation of the making of iron.

To the spectator who suddenly comes upon this gaunt assemblage of abodes, and forms a gloomy picture of what life must be like in them, it is an actual consolation to know that many of the dwellers in the place have as deeply rooted an attachment to it as though it were a beautiful village. There are people living in these hard-looking, shabby, ugly streets who have been there for many years, and more than one who has left it has actually pined to be back again. It is not, after all, every man or woman who is susceptible to scenery and to the outward aspect of the world round him; there are many who are nourished by human intercourse rather than by natural beauty. And such as these can find companionship and stimulus in the population we are describing; for many of the dwellers in these cottages, as those know who have frequented them, have a veritable keen zest in existence, a fund of human sympathy, and a spirit of enterprise as applied to mental as well as physical toil.

From the point of view of the workmen there are even advantages in living in the centre of the works in these desolate surroundings; it means that instead of having a cold windy walk to and from their work every day they are on the spot. For it must be a substantial addition to the hardships of life to have every morning and every night in the winter to cross that cold wind-swept river with the additional walk on either side. It is counted a privilege, therefore, to live in this strange wild settlement, and since the number of cottages is limited, it means that the workmen who live there have a claim of some personal nature, usually that of a long term of service. That claim of long service is one which many of them are able to put forth.

Out of 260 odd,

    27 had been at the works over 2 years.

42	,,	,,	,,	5	,,
25	,,	,,	,,	10	,,
24	,,	,,	,,	15	,,
6	,,	,,	,,	20	,,
11	,,	,,	,,	25	,,
43	,,	,,	,,	30	,,
33	,,	,,	,, all their lives.		

    46, time not mentioned, although six out of this number had been there 'many' years.

Of those who had been there over thirty years, the periods of service varied from thirty-one years to forty years, only nine of the men living on the north side of the river having been, under two years, at the works we are describing.

Out of 585 workmen living in Middlesbrough,

    41 had worked for the same employers under 2 years.

104	,,	,,	,,	over 2	,,
98	,,	,,	,,	,, 5	,,
45	,,	,,	,,	,, 10	,,
45	,,	,,	,,	,, 15	,,
11	,,	,,	,,	,, 20	,,
2	,,	,,	,,	,, 25	,,
12	,,	,,	,,	,, 30	,,
40	,,	,,	,,	all their lives.	

Of those who have been at the same works **over**

thirty years, two have been there forty years and
one forty-two years.

The greater number of the workmen who live on
the south side of the river, in the town, have every
day to walk about a mile or more to the ferry,
according to the place of their abode, to and from
their work, a walk hardly less dreary than the one
on the north side. It leads through one of the worst
quarters of the town, along a wide but wretched
street, looking the more wretched for its width—a
street of pawnbrokers, of second-hand shops, of
cheap night shelters, of openings into crowded
yards; a street in which one feels increasingly as
one approaches the ferry that riverside quality into
which the quarters of every town that lie near the
wharves and banks always seem to deteriorate.
There is something in the intercourse of sailors
from other ports who come and go, nomadic, un-
vouched for, who appear and disappear, with no
responsibility for their words or their deeds, that
seems to bring to the whole world a kinship of law-
lessness and disorder. But happily the road on
either side of the river, traversed by the workman
every day, and in many cases familiar to him since
his childhood, does not probably appear to him as it
does to the spectator who sees it for the first time.
For to us all, the salient points of any spectacle are
the unaccustomed, and all that is salient to us in
this progress has long since ceased to arrest his

attention. It is a matter of common experience that we wonder at a novel sight the first time we see it, we accept it the second time, we overlook it the third; until the deposit of daily habit has so obliterated the aspect of our daily surroundings that they cease to produce upon us any definite impression.

CHAPTER II

THE PROCESS OF IRONMAKING

THE brief description attempted in this chapter of the process of making iron does not profess to be anything but a crude elementary sketch, designed to give some idea of the daily work of the men employed at the ironworks, and the conditions under which it is accomplished.

The materials required are, roughly speaking, ironstone, coke, and limestone, which in the district in question are all found comparatively near at hand. The town and works we are describing have the further great advantage of standing, as has been already said, on the banks of a big and navigable river, at a few miles from the place where it falls into the North Sea.

The iron ore, or ironstone, is brought to the works from the mines in the Cleveland Hills, the limestone from the quarries in the Pennine Hills, and the fuel from the coal-mines in the county of Durham, either in the form of coal or of coal already transformed into coke.

It does not come within my scheme to enter into any description of the work either of the collier, who, working underground in the narrow passage cut in the seam of coal, sometimes so low that he is not able to stand upright, loosens the coal with a pick, or of the iron-miner who, working underground in a gallery 10 feet high, drills holes for the reception of gunpowder into the ironstone, which is afterwards blown into fragments. I take up the work at the stage where the iron and coal have been already 'won,' according to the technical phrase, and have been brought from the mines to the works.

The lumps of iron ore, looking like rough pieces of ordinary greenish stone, averaging about 10 inches in diameter, are charged on trucks at the mines and conveyed to the works. Arrived at the works, these trucks, containing severally the ironstone, the coal, the coke, and the limestone, are run along the lines laid on the top of two long iron platforms called 'gantries,' about 60 feet high from the ground, in each of which are at intervals apertures through which, by opening the movable bottom of the trucks, the materials are emptied into their various destinations. Beneath the gantry is a series of big partitions of brick, called 'bunkers,' practically forming store-cupboards, open at the back and front, in which are heaped the stores of coke and limestone to use in the furnace. There is also a certain amount of ironstone stored in case the supply coming from the mines in the trucks should run short, for the furnace must not cease burning day or night, and must have its constant and continuous supply.

Underneath the other gantry, beneath which begins the actual process of the manufacture of iron, stand at certain intervals huge kilns in a row, looking like black round towers made of iron. These kilns are about 45 feet high and 21 feet in diameter, tapering to 13 feet 6 inches, joined together at the top by the gantry, in which the tops of the kilns form a series of large round openings.

In these kilns the ironstone is calcined, in order to drive off the moisture and carbonic acid contained in it. The fire in the first instance is kindled at the bottom of the kiln by the usual methods, after which the ironstone is dropped into the kiln from the gantry above through the opened bottom of the truck, brought over the hole at the top of the kiln. The kiln is fed with coal as often as necessary. The fire gradually rises. Once alight, it is not allowed to go out, but continues burning night and day unless it is necessary to extinguish it and relight it for repairs, or from some accident. All round the bottom of the kiln there are huge iron shutters, which let down, so that the ironstone, when sufficiently roasted, can be drawn out through these, taken on barrows, and wheeled away to the furnace. It is now, after having been calcined, of a dull red. These barrows, together with other barrows containing, some, coke, and others limestone (the limestone at the first cursory glance looks not unlike the ironstone before calcination, but has a more decided blue tinge, the ironstone being green), are then placed upon a lift which goes up to the top of the blast-furnace, into which the contents of the barrows are from thence emptied.

The blast-furnace is, to look at, a huge round black tower of iron, 80 feet high and 30 to 40 feet across. It is filled from the top, where there is a great circular opening like the mouth of an immense funnel, but closed by a ' bell,' which can be lowered, on to which the charge of the furnace is thrown. The charge put into the furnace at one time is six barrows of ironstone, six of coke, three of limestone. This charge is called a ' round.' The coke is put in for fuel, the limestone is put in to serve as what is called a ' flux '—that is, a material which in combination with another makes a fusible compound. For the iron-stone contains silica, or sand, and alumina, or clay, which are difficult to melt : the combination with the limestone makes them fusible. The mass of heated material is constantly sinking down through the furnace, and being drawn out at the bottom in the form of pig-iron and slag.

About 6 to 8 feet from the bottom of the furnace a strong blast of air, playing the part of a gigantic bellows in fanning the flames inside the furnace, is driven in by machinery. This blast is driven in hot in order to avoid unnecessary cooling of the interior of the furnace. The gases generated by the process, including 23 per cent. of burnable gas, which can be seen blowing off at the top in a stiff whistling flame, are constantly escaping. By a skilful arrangement these gases are now conducted through a huge flue, and thence distributed to the tubes which lead them through the 'stoves' (tall black cylinders about 60 feet high and 21 feet across, made of honeycombed brickwork, also like round black towers, but smaller than the furnaces) and so beneath the boilers which drive the engines propelling the blast into the furnace.

By this ingenious arrangement of circuit, therefore, the gases engendered by the furnace return to heat the blast which blows it, since the stoves become white-hot by the passage of the hot gases, and the blast blown into the furnace is heated by passing through the stoves. The gases are directed into the stoves alternately. They pass through one stove and heat it white-hot, then the gas is shut off and the air is blown through the stove in the reverse direction, cooling it and becoming heated itself—after which the blast is shut off, and the gas, which in the meantime has been heating another stove (into which the air is now again directed), again passes through and reheats the first. At the bottom of the furnace, when, a given time having been allowed, the moment has come to draw the iron out, a hole is made, and the red-hot molten iron flows out into sand-moulds prepared for it, in which it cools in the shape of short iron logs.   These logs are what is called pig-iron.

The molten iron destined to make steel, instead of flowing into the sand-beds, is drawn off into a huge iron vessel called the ' ladle,' and of this I will only say briefly that, the impurities which make the iron too brittle and too fusible for the purposes for which steel is required having been subsequently removed by exposing it to a very high temperature, it is then cast into large blocks, called ingots, of what has now become steel.   I do not propose to attempt any description of the process by which these ingots are rolled out red-hot into great bars, billets, angles, etc. From the human point of view the surroundings and the conditions of life are practically the same at the steelworks as at the ironworks.

This is a bald summary of the process of the making of iron.   Let us now consider it in more detail, with the ' job ' of each man, as he would himself call it, and the conditions under which he works.   The workmen comprise furnace-keepers, slaggers, chargers, mine-fillers, brakesmen, weighers, coke-tippers, helpers, slag-tippers, gantrymen, scarrers, metal - carriers, boiler-men, the engine-men, the men who look after the pumps, the cranes, the stoves, the labourers who help in various departments: the men who work in the ' shops '—*i.e.*, workshops—joiners, moulders, pattern-makers, who shape the various pieces of wood from which the moulder makes the moulds in which the iron will be cast for making pieces of machinery, and what is called the sailor fitter gang, practically odd men, some of whom have in reality been sailors, and who do any rigging up of anything incidental that happens to be necessary. The best paid of these men get from £2 to £3 per week, the lowest 19s. 6d. The majority are in receipt of from 25s. to 38s. per week.

The working day, since the eight-hours day was agreed upon between masters and men, is now divided into three ' shifts ' or spells of work of eight hours each. The first shift is from 6 a.m. to 2 p.m., the second from 2 p.m. to 10 p.m, the third from 10 p.m. to 6 a.m. On the Sunday, the odd day in each week, the men who have been during the preceding six days on the morning shift (from 6 a.m to 2 p.m.) work sixteen hours on end—that is, from 6 a.m. to 10 p.m.—and then begin the next day at 2 p.m. and go on till 10 p.m. Every third week, therefore, they work on this long shift through the Sunday.

This distribution of work applies to those workmen whose job cannot be left for a moment—*i.e.*, those who are concerned actually in dealing with the ironstone from the moment that it arrives at the works until, converted into iron, it is carried away. The hours of the others are not divided in the same way. The metal-carriers work from 6 a.m. until their work is done; the slag-tippers work according to the tide, at both the low tides; the crane-drivers according to the tide and the shipping. Some of the locomotive drivers have the eight-hours shift; some work from 6 a.m. to 5 p.m. The labourers work from 6 a.m. to 5 p.m.

The blast-furnaces are never allowed to go out night or day, week-days or Sundays, on account of the difficulty, trouble, damage to the furnaces, and great expense of relighting them. During the Durham coal strike in 1892, when many of the works had to be laid idle, the fires were banked up in the furnaces that they might not go out; and thus kept up on this colossal scale, they remained alight for three months. This, though better than if they had been allowed to go out and had had to be relighted, was very detrimental to the furnaces, and necessitated time, trouble, and repairs to get them into their normal condition again.

The men who work on the 'gantry'—that is, the tall platform along which the trucks arrive containing the ore, the coal, the coke, and the limestone— are called 'gantrymen.' A gantryman has from £2 to £2 12s. 6d. a week, an income equal to that of many a curate, and that of many a junior clerk in a private, or even a Government, office. For this he works either eight hours a day or six hours a day, according to the shift, for the gantryman only has two shifts in the day, and not three, either from 6 a.m. to 2 p.m., or from 4 p.m. to 10 p.m. The daily work of the gantryman, who stands on the platform at the top of the kiln, is to superintend the calcining of the ironstone—*i.e.*, to feed with coal the kiln into which the ironstone has been 'tipped.' The coal is tipped, not immediately into the kiln, but on to broad, flat sheets of iron on either side of the opening on to which the coal has been tipped from the truck. The gantryman at intervals shovels the coal into the kiln on to the top of the ironstone, using his own judgment how much he puts in, and how often.

The shift from 6 a.m. to 2 p.m. is the hardest, as at 6 a.m. the kiln has been left since ten the night before, so that there is a great deal to do. Then by two o'clock it is so full that it can be left until four. This second shift has not so much to do as the first, as it has only two hours interval to make up. Each of the eight-hours men is supposed to have two quarters of an hour in which to have some food. This food—his 'bait'—as he would himself call it—he takes up with him. It is usually something that can be carried in a can, and kept hot in the little 'cabins,' of which one is allotted to each particular gang of men working at the same job —comfortless little sheds enough, filled with dust and dirt and the smell of the gases. For a man working on the eight-hours shift, of course, it is possible—as far, that is, as time is concerned—whatever shift the man is on, for him to have a square meal at some reasonable hour. The men who go off at 2 p.m. can have their dinner at home in the afternoon ; those who come at 2 p.m. can dine at home before leaving. A man who comes to his work in the evening has had his meal at home at suitable intervals during the day, if his wife is competent.

But none of the eight-hours men when at the works can leave their job for a square meal, although they may have short spells off in which to sit down either in the cabin or in the place where they are working and have a snack. They are not for one moment free from the need of that incessant vigilance required from all the workers who share the responsibility of controlling the huge forces they are daring to use.

The gantryman, not being close to the great heat of the furnace, has not to encounter in the winter the same sudden and violent changes of temperature as those which the men standing round the foot of the furnace must endure, when they pass from the almost unbearable heat of their immediate surroundings to the biting cold of a few yards off. But still, the work of the gantryman is a task absolutely shelterless, and in the sharp airs of the north-east coast the winds that blow across the high platform pierce like a dagger ; added to which they bring the burning fumes

which blow across from the kilns and the furnaces,
the choking sulphurous fumes which are constantly
blowing down the throats of the men exposed to
them.

And besides the fumes and the gases, every breath
of wind at the ironworks carries dust with it, whirl-
ing through the air in a wind, dropping through it
in a calm, covering the ground, filling the cabins,
settling on the clothes of those who are within reach,
filling their eyes and their mouths, covering their
hands and their faces. The calcined ironstone
sends forth red dust, the smoke from the chimneys
and furnaces is deposited in white dust, the smoke
from the steel-rolling mills falls in black dust : and,
most constant difficulty of all, the gases escaping
from the furnaces are charged with a fine, impalpable
brownish dust, which is shed everywhere, on everything, which clogs the interior of the stoves and of
the flues, and whose encroachments have to be constantly fought against. One of the most repellent
phenomena at the ironworks to the onlooker is the
process of expelling the dust from the stoves, for
which purpose the valves of the stove are closed, the
stove is filled with air at high pressure, and then one
of the valves is opéned and the air is forcibly ex-
pelled. A great cloud of red dust rushes out with a
roar, covering everything and everybody who stands
within reach, with so intolerable a noise and effluvium
that it makes itself felt even amidst the incessant

reverberation, the constant smells, dust, deposits, that surround the stoves and the furnaces. That strange, grim street formed by the kilns, the furnaces, and the bunkers, darkened by the iron platforms overhead between the kilns and the gantry, a street in which everything is a dull red, is the very heart of the works, the very stronghold of the making of iron, a place unceasingly filled by glare, and clanging, and vapours, from morning till night and from night till morning.

The material tipped by the gantrymen into the kilns is gradually, as it sinks down, drawn off at the bottom through the 'hoppers,' which can be opened or closed as required by the huge iron shutters before described. A hopper, be it said, is practically any vessel or receptacle with a hole at the bottom and sloping sides. Every now and again, if the kiln is too hot, the ironstone begins to fuse and the pieces stick together, making a lump too big to get through the inside of the kiln into the hopper. The lump is called a 'scarr,' and it then has to be broken into bits with an iron rod from the outside by a man called the 'scarrer,' who stands on a movable wooden trestle, about the height of the bottom of the hopper, and breaks up the lump by thrusting his bar into the opening until the lump is small enough to come through. The opening into the kiln above the hopper is called the 'eye.' The scarrer earns from 30s. to £2 a week; he spends most of the eight-hours day standing at the bottom of the kiln, the iron rod in his hand, ready to thrust it into the kiln whenever the obstacle shows itself. By him is standing another man ready to add his weight to the thrust if the strength of the first one is not enough to deal with the obstacle. For whatever operation is being carried on at the ironworks, there are always a number of men standing round in a state of watchful concentration, their attention on the alert, ready to lend a hand in a case of emergency. The spectator receives an overpowering impression of what that watchfulness needs to be, of what sudden necessities may arise, of what may be the deadly effect of some swift, dangerous variation, some unexpected development in the formidable material which the men are handling.

The 'scarrs' having been broken, and the calcined ironstone, now of a dull red colour, drawn out from the hopper into shallow iron barrows, it is wheeled from the kiln to the lift of the blast-furnace, 20 to 30 yards off, by a man called a 'mine-filler,' who does this ten to a dozen times during the hour. On the way to the furnace he stops to have the weight of his load checked outside the 'weigh cabin,' a little dark, dusty shed, outside the window of which there is a platform like that used for weighing luggage at a station. The men get so well accustomed to judging the 'burden' required—that is, the weight of ironstone or limestone that ought to be in the barrow—

that they generally hit off almost exactly before it is weighed the required quantity of rough pieces of stone piled up in it.   The mine-filler is paid 42s. a week.

The weight being ascertained to be correct, the weighman puts down a straight stroke for each barrow that passes him, and after every four draws a horizontal line through the four strokes, each of these groups of lines therefore representing five barrows.   The man in the weigh-cabin receives about £1 per week.   He is generally one of the older men, obliged to accept as his strength declines a lighter and less remunerative form of work. ` His job is to sit for the eight hours in which he is on duty in his little cabin, sheltered, at any rate, from the weather, but not from the smells, the noise, and the dust of the blast-furnaces close to which his cabin stands.

The mine-filler, the weight of his barrow being checked and found correct, then wheels it to the lift, a square, unenclosed platform which goes up and down to the top of the furnace and back again between four huge supports.

In the place we are describing there are first two furnaces and then a lift, then two furnaces again, then a lift, and so on ; the same lift always serves the same pair of furnaces.   The six barrows of iron-stone, three of limestone, and six of coke, constituting the ' round' already described, are placed on the lift and hoisted to the top of the furnace, where they are taken off by the chargers, and wheeled close to the big round aperture at the top of the blast-furnace. These men are standing on a platform about 10 feet wide that runs round this aperture, and is guarded on the outside by a railing. That aperture is closed by a huge lid, a bell and hopper in this shape : the centre edges of a hopper (which in this case is the shape of a deep saucer or shallow bowl with its middle out) rest upon the sides of a shallow cone called the ' bell.' The charge is lowered on to the top of the bell, on which the inner edge of the hopper is resting ; the bell is then lowered, leaving, therefore, an opening, and the charge, no longer kept in place by the sides of the bell, slides down them into the furnace below. The bell is then rehoisted into position. The sudden jet of vivid flame from the top of the furnace, so familiar to those who live in the neighbourhood of ironworks, is produced at the moment of charging. The chargers have from 30s. to £2 per week. The work of the charger is arduous and trying to the health. Men with susceptible lungs are apt to be much affected by the combination of the rapid breathing necessitated by handling the heavy barrows and the fumes inhaled with every panting step.

The temperature inside the furnace is about 3,000° F. This great mass of combustion is being perpetually fanned by the hot blast, heated, as has

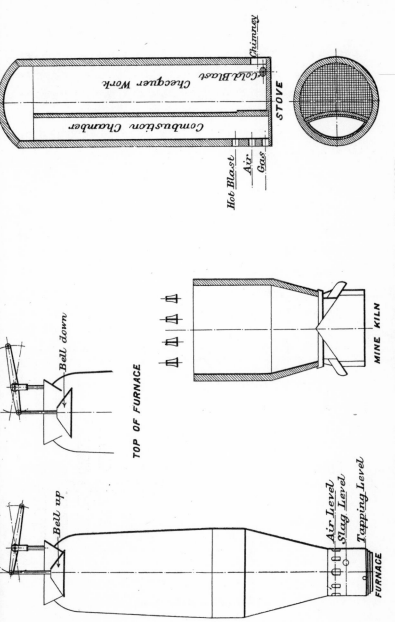

STOVE.

Hot Blast
Air
Gas
Combustion Chamber
Chequer Work
Cold Blast
Chimney

Bell up
Bell down
TOP OF FURNACE

MINE KILN

Air Level
Slag Level
Tapping Level
FURNACE

SECTIONS OF FURNACE, MINE KILN, AND STOVE.

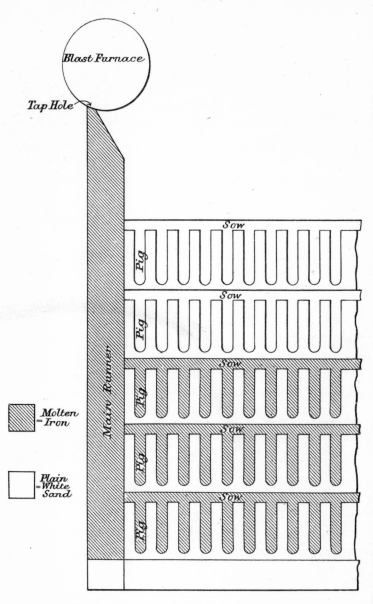

Blast Furnace

Tap Hole

Sow

Pig

Sow

Pig

Sow

Pig

Sow

Pig

Sow

Pig

Main Runner

▨ Molten
= Iron

☐ Plain
= White
Sand

GROUND-PLAN OF PORTION OF 'PIG-BEDS.'

been before described, by passing through the stoves
made hot by the passage of the gas, and incessantly
driven into the furnaces at the top of the 'hearth,'
just above the slag opening, by blowing-engines
moved by machinery. The iron, being heavier,
gradually sinks to the bottom, and the slag (which
may be called an artificial lava, and is practically the
scum or dross of the iron) rises, floating on the top
of the iron, just below the place where the blast is
blown in. The materials put into the furnace
gradually sink down as those which preceded them
are consumed, and the metal is separated from the
dross with which it is associated in the stone. As
they sink, more material is put in above, so that the
level of them at the top is always within a few feet
of the bell. The heavier iron falls molten to the
bottom, while the dross floats molten on its surface.
As the process continues, the mass of liquid iron
rises, as well as the slag or dross, until the latter
flows out of the furnace through a hole provided for
the purpose. Four times in the twenty-four hours
the furnace is 'tapped'; a hole is made at the bottom
of the furnace, and the iron is allowed to run out.
This is an important moment in the process of
manufacture. To say that a hole is made in the
furnace for the iron to run out sounds simple enough:
but a stream of molten iron cannot be drawn off like
water, and time after time the tapping of the furnace,
accomplished by breaking by main force through the

piece of fireclay, which, having been thrust cold into the red-hot aperture after the last casting, has baked and hardened in the opening to a solid mass, is a strenuous encounter with a potent and deadly enemy. To be a 'furnace-keeper,' and responsible for the furnace being in absolute working order, is one of the most responsible posts at the works. The keeper gets from £2 10s. to £3 per week.

A great square platform, the 'pig-bed,' of firm moist sand, dug from the river-bed, extends 10 feet from the ground in front of each furnace, at a level of about 1 foot below the bottom of it. In this, before each casting, the channels are prepared into which the molten stream is to run, by the helpers of the furnace-keeper. These channels consist of one long main channel, 16 inches wide and 10 inches deep, at right angles to which are other channels varying in number, generally about 16 feet long; between these, parallel to the main channel, are rows of shorter channels, also varying in number according to the size of the pig-bed; there may be twenty-four, or there may be even as many as thirty-six. These shorter channels are for the pieces of iron which when cast will be known as 'pigs'; the transverse and longer channels are called the 'sows.' The area of the sandy platform having been made tolerably smooth and level again after the last casting, the men take first of all a long piece of wood the size that the sows are to be, with an iron ring at each end of it for facility of handling, and put it across the platform where the first sow is to be. They then take a number of short oblong blocks of wood, the shape of the pigs, and drop them rapidly one after another all along at right angles to the sow, with about 3 inches of sand between them ; other men standing ready with spades then throw sand into the interstices between the wooden blocks, forming a partition between each. When the row is finished, therefore, each block of wood is lying practically in a little rectangular sandy hollow. The long transverse piece, the sow, is then lifted up by two men holding the iron ring at either end and thrown across to others, who put it across at the bottom of the shorter ones, and repeat the operation. Then all the shorter wooden pieces are taken up by the ring at the end of them in the same way (except that they can be lifted by one man) and thrown across to the others standing below the next sow, who repeat the operation, as with the sow of dropping them into position and filling up the interstices with sand. Into the first row of moulds, now left empty, a man holding a wooden instrument, a thin piece of wood fastened transversely at the bottom of a long handle, like a broom ending in wood instead of bristles, goes rapidly along the channels, flattening them at the bottom ; after which he is followed by another man with an instrument which has at the end of its handle a cross-piece of iron with raised letters on it, with which he in his turn goes along the channel of pigs, stamping the name of the brand at the bottom of them. And so admirably adapted is the river sand for the purpose of a mould, especially after several castings have been run over it, thoroughly drying it and burning the lime out, that this stamping by hand by a mould with raised letters, firmly pressed into each channel in turn, is enough.

To the outsider, indeed, part of the absorbing interest of watching the manufacture of iron is that in this country, at any rate, it is all done by human hands, and not by machinery. From the moment when the ironstone is lifted off the trucks, then dropped into the kilns, afterwards taken to the furnace, and then drawn out of it, it has not been handled by any other means than the arms of powerful men, whose strength and vigilance are constantly strained almost to breaking-point. It cannot be too often repeated what the risk is of dealing with a thing which you encounter only in terms of liquid fire. The path of the ironworker is literally strewn with danger, for as he walks along, the innocent-looking fragment, no longer glowing, may be a piece of hot iron of which the touch, if he stepped upon it, is enough to cripple him ; one splash of the molten stream may blind him ; if he were to stumble as he walks along the edge of that sandy platform where the iron is bubbling and rushing into the moulds he would never get up again. The men move about among these surroundings with the reckless — often too reckless—unconcern of long habit. You may see as you pass a man standing engaged in thickening the end of a bar of iron by leisurely twirling it round and round in a vessel full of red-hot slag, of which he will then allow a portion to cool on it, and doing it as calmly as though he were stirring round a pan of water.

When the moment comes to open the furnace for the next casting, the requisite time having elapsed since the last, a hole is drilled through the bottom of the 'hearth' in the solid piece of clay with which the furnace was closed the last time it was tapped. This is done by a great iron bar, held by three men, being thrust again and again against the clay, clearing away the loose rubbish and beginning to make the hole, against which it is then held in position by two men, while two others deal alternate blows on it with big hammers. It is done as quickly and regularly as though by machinery, the top of the crowbar hit fair and square with a clang at every blow, and the men who are striking are surrounded by a group of others waiting. These last are slaggers, mine-fillers, and others, come away from their special job to help in this one. They all stand intently concentrated on the moment when the clay shall finally yield. When that moment has come, the bar, with a mighty effort, is withdrawn, and the blazing stream rushes out in foaming and leaping red-hot waves through the opening, into the deep main channel prepared for it. The heat to those standing on the bank of that molten river is almost unbearable, the glare hardly to be endured. The onlooker is half blinded, half choked, as the flood rushes fiercely past him.

The men who are watching are on the alert, and with the iron bars thickened at the end by a lump of slag, the stream is controlled. It goes rushing down the whole length of the main channel to the last set of moulds, into which it is allowed to flow by having the barrier of sand which bars the access from the main channel broken down. So have we all of us in our childhood dug channels in the sand for the incoming tide, dammed them up, and then dashed one open again with our spades as the first encroaching wave came foaming along. The moment one of these sets of channels is filled, the man whose job it is, working gradually up nearer the furnace, breaks down the barrier into the one next above it, and the stream dashes into that, while at the same time the iron in the main channel is stopped from rushing below that point by having a big spade-like implement stuck across it. One channel after another becomes a quivering red mass of liquid fire as the hot iron flows into it, taking the shape of the mould as it cools. As each set of channels is filled, sand is rapidly thrown over the surface of the molten iron by men with spades, and after it has somewhat cooled cold water is thrown over it by a hose.

During this time the aperture into the furnace has been closed again by filling it with lumps of fireclay. A man, taking one of these lumps—or more, if necessary, in succession—and kneading it with his hands into a rough ball, stoops and hurls it with all his might straight into the funnel-shaped opening at the end of which the red-hot iron can be seen bubbling, the throw being instantly followed by a mighty thrust from a huge bar of iron, wielded by five men, who stand ready as each lump is thrown in to push it home. When the hole is finally filled the fireclay hardens in it, baked to a solid mass, and the aperture is sealed, to be reopened six hours later, and the whole process gone over again. Sometimes the closing of the aperture is effected in three minutes; sometimes, not often, it has taken nearly as long as an hour.

At this stage, of course, each sow is still joined to a row of pigs. As soon as they get rather cooler, but while still nearly red-hot, the little short logs, the pigs, are broken off from the main runner, the sow, which is done simply by putting an iron bar underneath them and prizing them up, for in this state the metal is very brittle. They are then lifted by pincers and put up on end leaning against one another at the top, and again watered by a hose to make them cool more rapidly. Then finally when it is possible to handle them, they are taken up and carried one at a time by the metal-carriers to the trucks which convey the metal either to the stock-yard, the place where the stores of pig-iron are kept piled one on the top of the other ready for transport by sea or land, or else directly to some ship which is going to convey the iron to the purchaser. About eighteen of these pigs go to the ton : that is, each of them weighs about 9 stone. A man lifts one of them at a time, holding it by the two ends with the middle resting on his leg. The metal-carriers do not work on the eight-hours shift : they work until the job they have in hand is done. That is, if a given quantity of pig-iron is going to be taken away on a vessel they work until it is all carried down to the bank, or if it is going first of all to stand in the stock-yard, it is carried to the stock-yard to be piled up ; and the men are paid according to the amount of iron they carry.

The loading and unloading is done by 'stevedores,' or stowers, at the river-side. A stevedore gang gets sometimes 20s. or over per 100 tons, which is divided among the gang.

The slag — the dross of the iron, as described above—flows either into ladles or into round vessels called 'bogies.' The slag flowing into the latter hardens in cooling into what are commonly called slag 'balls,' though, of course, they are not balls, but short broad cylinders. These ladles are then taken along a railway-line to various places, where they are 'tipped'—that is, emptied out on to heaps, which eventually form embankments. A frequent sight on a winter's night, one of the sights by which one 'visualizes' the ironworks, is a slag ball bursting as it is tipped, and flaming up into a mass of flying fragments as it rolls down. The sides of these embankments, with the brittle jagged edges of the slag, sharp as glass, are the places where the children of the ironworkers clamber up and down, as well content, apparently, as the more fortunate children who are rolling down a grassy slope. But a slip in these jagged edges means something very different from a slide and a roll down a grassy slope; and the children playing on the slag 'tip' are face to face with a daily danger, which grows more dangerous, and not less, as familiarity with it makes them more heedless.

These long grey bare headlands do at last, with the passing of time, become gradually clothed with green, but not till they have stood for many years. It is but a parody of scenery, at best, amongst which the children of the ironworks grow up. The world of the ironworks is one in which there are constant suggestions of the ordinary operations of life raised to some strange, monstrous power, in which the land runs, not with water, but with fire, where the labourer leaning on his spade is going to dig, not in fresh, moist earth, but in a channel of molten flame ; where, instead of stacking the crops, he stacks iron too hot for him to handle ; where the tools laid out ready for his use are huge iron bars 10 feet long or more, taking several men to wield them. The onlooker, whose centre of activity lies among surroundings different from these, walks with wonder and mis-giving through the lurid, reverberating works, seeing danger at every turn, and shudders at what seems to him the lot of the worker among such grim surroundings as these. But there is many a man employed in the works to whom these surroundings are even congenial, to whom the world coloured in black and flame-colour is a world he knows and understands, and that he misses when he is away from it. And there must be hundreds and thousands of people earning their livelihood in other ways, whose actual working hours are passed in a setting that will seem to many of us still less enviable : who, adding up figures or copying letters, see nothing but the walls of one small room round them for eight hours every day. For the actual nature of the occupation of the various branches of ironmaking may appear to some of us, given the requisite strength and the requisite health, to be preferable to many other of the callings which might presumably have been open to the people engaged in it, or, indeed, to those open to other classes of society who spend their lives in sedentary anxiety.

Many of the people who have to sit in a stuffy room and pore over pages in a bad light, would probably find their nerves, physical condition, digestion, and general outlook on life entirely different if they were, for instance, making trenches in the sand, taking out moulds, and throwing them to the next man to do the same.   There is, no doubt, monotony in most occupations; but it is conceivable that the monotony of the work conducted in movement is not so penetrating as that which is conducted in immobility.   It has not, at any rate, the deadly uniformity of being indoors, and even in England there are many days in the year when the weather is fine, and when the vapours from molten iron, ill-smelling though they be, cannot entirely sweep away the sweet breezes from the river.   There are many of the men, no doubt, who look sullen at their work, many who look discontented; but so there are in all callings.   In any Government office that you can name, I believe that there is as large a proportion of sullen and discontented faces as there is at the ironworks; indeed, there are probably more, because there have been more possibilities to choose from, and the area of disappointment is therefore greater. People differ in their conceptions and their definitions of happiness.   A recent paraphrase of Aristotle has defined it thus: 'Happiness consists in living the best life that your powers command in the best way that your circumstances permit.'   The man at the ironworks, who by his character and aptitude, is safe to have regular employment, whose health is good, who has a sufficient wage to have a margin, and a wife, that he cares about, competent enough to administer it to the best advantage, who has a comfortable home and children of whom he is fond, seems to me to have as good a chance of happiness as most of his fellow-creatures—and there is fortunately many a man at the works to whom this description can apply.

CHAPTER III

THE EXPENDITURE OF THE WORKMAN

In considering the pecuniary circumstances of the ironworkers we are describing, we should not expect to find many of them in absolute poverty, since we are dealing only with those who are employed at the ironworks, and assumed, therefore, all to be at work and in receipt of regular wages.

The dire need of the people who cannot find employment has been so ever-present to our minds in recent days that we are apt to believe that once employment is secured, once what seems like regular work is obtained, all must be well, so long as the workman is steady and knows how to manage his money. But when we take this rosy view we postulate a great deal, and we forget how terribly near the margin of disaster the man, even the thrifty man, walks, who has, in ordinary normal conditions, but just enough to keep himself on. The spectre of illness and disability is always confronting the working-man ; the possibility of being from one day to the other plunged into actual want is always confronting his family.

The wages of the ironworkers, broadly speaking, range from 18s. to 80s. per week. There are some boys, not many, employed at the works who receive less.

Out of 1,270 people paid in a given week, 23 (these were boys only) received under 10s. per week, 50 more boys received under 20s., 96 men, mostly labourers, received under 20s., 398 received from between 20s. and 30s., 410 between 30s. and 40s., 235 between 40s. and 60s., 58 between 60s. and 80s., and 4 over 80s.

When a man first begins work, he is not paid until the end of the second week, instead of being paid at the end of the first. The employers, therefore, always owe him a week's wages, which is practically a surety for them that he will not disappear at a moment's notice.

These wages can be further supplemented in other ways, such as working overtime, etc. The labourers in receipt of 3s. 3d. per day often get a good deal more by working in the place of absentees, also by discharging ore from steamers and bricks from trucks, etc. The wages of the blast-furnace men vary with the tonnage of iron made, according to the sliding scale in operation at the time. The weekly output of a blast-furnace is estimated at 500 tons, and over this amount the men get a percentage on it.

Some of the houses, about one-third, take in lodgers. The lodgers pay from 12s. to 15s. per week for lodgings, board, and washing, and the woman generally is supposed to make a profit of between 2s. and 3s. on each person. Of course, with this substantial addition to the income, there may be sometimes increased discomfort in the house, since there is less room, and more work for the wife. The lodger is often on excellent terms with the family, and is treated as one of them.

These wages seem high, no doubt, and many of the men enumerated are well off and prosperous; but not necessarily always those with the higher incomes. Some of the most comfortable homes are those where the man is in receipt of about 30s. more or less, and has a wife and several children to provide for out of it. It is a matter of common experience that the household that is managed admirably when it is necessary to consider every penny has with a sudden increase of wages fallen on evil days; the desperate tense strain of incessant economy being a little relaxed, the unfamiliar possibility of spending has been allowed to go too far.

The time when existence seems to press most hardly is during the first twelve or fourteen years after marriage, when there is usually a family of young children, who have to be provided for and who cannot earn; and the wife is constantly, before and after every birth, in a condition in which she cannot fulfil her duties with efficiency. It is not until later that the husband's wages are supplemented by odds and ends of work on the part of the wife, when she has a daughter old enough to leave in charge of the house, and also by earnings of one kind and another from sons and daughters employed in other callings. If the household succeeds in arriving without mishap at this stage, the outlook is better. But the income during these first years of married life being almost entirely dependent upon the health and physical condition of the one bread-winner, any illness or accident affecting him at once plunges them into difficulties. The men are constantly breaking down in health, either from conditions inherent in the work —the noxious fumes, the violent alternations of temperature, to which they are necessarily exposed—or simply because they were not strong enough at the beginning to follow an occupation which, of all others, requires great physical strength. The man is thrown out of work: and then come the dreaded lean weeks in which the wages, which may just have sufficed before, when carefully administered, cease altogether, at a time when he is ill and probably requires some different, better, and therefore more expensive, sort of food.

We may call those of the ironworkers absolutely poor who have actually not money enough to buy what are called the necessities of life—food, drink, fuel, and clothing, and a house over their heads; and those relatively poor who would have enough if they did not spend some of their means on something else, or if they had not had some unforeseen call, such as illness, upon their resources. A keen observer who has spent his life in the town I am describing, gives it as his opinion that 'there are not many of them who do not handle money enough to make them comfortable if it were rightly used.' And this seems to me to put the case accurately. Most of them do have the opportunity of handling money enough, but have not skill or self-control to handle it to the best advantage. To discuss the pecuniary means of the working classes, or, indeed, of any other class, without almost always allowing for leakage and waste is to start on a misleading basis. Out of 900 houses carefully investigated, 125, in round numbers, were found to be absolutely poor. The people living in them never have enough to spend on food to keep themselves sufficiently nourished, enough to spend on clothes to be able to protect their bodies adequately, enough to spend on their houses, to acquire a moderate degree of comfort. One hundred and seventy-five more were so near the poverty-line that they are constantly passing over it. That is, the life of a third of these workers whom we are considering is an unending struggle from day to day to keep abreast of the most ordinary, the simplest, the essential needs. Four hundred were comfortably off as far as their means were concerned, though not more than half of them probably administered those means with wisdom and judgment; and 200 superior

well-educated homes were in quite easy and prosperous circumstances. The man who is in receipt
of £2 and upwards a week, who lives in a workman's
cottage, and does not need to dress other than as a
workman, does not need to be what the workmen
call a ' collar-and-tie man,' is obviously much better
off than many a clerk with the same amount. When
we find that a working-man and his household, who
have actually money enough, if it were not otherwise
spent, to buy that which would keep them in health
and comfort, are constantly in difficulties, it is the
custom to talk, and with some show of reason, of
improvidence and thriftlessness. That label is, no
doubt, easy to affix to their doors: but it is not in
every case that it is justified. Most of the people at
the ironworks are living under conditions in which
the slightest lapse from thrift and forethought is
necessarily conspicuous, and brings its immediate
consequences. These consequences may be the
result of a combination of many causes, although
it is, of course, possible, not to say probable, that in
most cases they have been caused by the lack of
thrift to which it is simpler to attribute them. There
are so many uncertain elements to deal with, such a
constant variation of possibility, dependent upon the
health, the temperament, the capacity, of both the
man and the woman concerned, that though we may
attempt to lay down general rules, we shall be
constantly misled in our judgment of individual

instances. When we say that what for one household is enough is for another absolutely not enough, we are simply restating what may be said of all human beings in any class. It is no good saying when we visit the H.'s or the J.'s, who are going to be sold up because they have not been able to meet their rent, that the A. B.'s have been able to keep abreast of circumstances; for it does not follow that because the A. B.'s have been able to achieve it the H.'s or the J.'s will be able to do so too.

It is, no doubt, happily possible to find among the ironworkers many homes that look bright and cheerful, and where the inmates are enjoying life; that is, as long as they are well. But any indisposition, any passing bodily ills, that to the well-to-do are uncomfortable enough in themselves, and are grumbled at as an interruption to the ordinary current of life, assume a more sinister aspect when physical discomfort and suffering are but a small part of the misfortune they entail, when there is not one penny to meet the extra expense by which alleviation would be bought, unless it is taken off something else which up to that moment has seemed essential. To those who are better off, to whom illness means added luxuries instead of curtailed necessities, whose economies when in health chiefly consist in determining which superfluity shall be rejected, and whose possible privations at any time attack their self-love more than their physical well-being, it may well be instructive to become acquainted with some of the household budgets of the poorer classes. I have been given the details of expenditure in various households, which make one marvel at the skill with which small funds are administered. At the same time, we must not assume that such skill reigns in all the households that we are considering. It is obvious that it is only the skilful and thrifty managers who administer their funds with such a nicety that they can put them down on paper. As the income increases it is much more difficult to acquire information; but the study of those who appear to be living on the minimum that is possible gives a basis on which to consider the finances of those whose income is a little larger. But in so doing let us not come to hasty conclusions: one may tabulate a man's income and expenditure, but it is not so easy to tabulate his skill in dealing with it, his temptations and his weaknesses; and all these have to be taken into account, as well as the number of shillings he receives. The ordinary compulsory expenditure of the workman includes the absolute necessities of house-rent, coal and wood, clothes, and locomotion, in a place where for many of the men the river lies between them and their work, and has to be crossed at a halfpenny a passage on a steam ferry-boat. It also includes fines imposed in certain cases at the works, such as being absent from work from any cause without notifying the fact, being found drunk while at work, etc. The amount stated for wages is the net amount received after certain authorized deductions have been made. In accordance with the Truck Act, by which it is illegal to levy any portion of the man's wages, even for purposes destined to his own advantage, unless at his own request or authorization, each man, on obtaining employment, goes to the timekeeper and signs a form authorizing the employers to make certain deductions from the wages. Each married man pays 3d. a week to the doctor, and each single man 2d., if, that is to say, they are in receipt of 18s. a week and upwards. The employers supplement this by a fixed remuneration to the doctor, who then attends the men and their families without any further charge. The rent, when the men are living in cottages belonging to the employers, is paid in the same way by deducting from the wages ; coals, when required, are also thus paid for. The workmen who live in the north-east of Yorkshire, and near the Durham coal-fields which supply this district, get coals on more advantageous terms than if they lived further away.

I give here six weekly budgets of the same household—father, mother, and girl of twelve, the man's income varying, as will be seen, from 18s. 6d. to 23s. 9d.

MRS. A. B.'S BUDGET.

First Week. Income 18s. 6d. Family Three.

	s.	d.		s.	d.
Rent	5	6	3 oz. tobacco ...	0	9
Coals	2	4	½ st. potatoes ...	0	3
Insurance ...	0	7	Onions	0	1
Clothing ...	1	0	Matches ...	0	1
Meat	1	6	Lamp oil ...	0	2
1 st. flour ...	1	5	Debt	0	3
¼ st. bread-meal	0	4½			
1 lb. butter ...	1	1		18	6
½ lb. lard ...	0	2½			
1 lb. bacon ...	0	9	SUMMARY.	s.	d.
4 lb. sugar ...	0	8	Rent	5	6
½ lb. tea ...	0	9	Insurance ...	0	7
Yeast	0	1	Coal	2	4
Milk	0	3	House	0	8
1 box Globe			Food	7	5
polish ...	0	1	Clothing ...	1	0
1 lb. soap ...	0	3	Tobacco ...	0	9
1 packet gold			Debt	0	3
dust	0	1		18	6

Second Week. Income 19s. 6d. Family Three.

	s.	d.		s.	d.
Rent	5	6	Clothing ...	1	0
Coals	2	2	Meat	1	7
Coke	0	3½	1 st. flour ...	1	5
Insurance ...	0	7	¼ st. bread-meal	0	4½

Second week—continued.

	s.	d.		s.	d.
1 lb. butter ...	1	2	Lamp oil ...	0	2
½ lb. lard ...	0	2½	Debt	1	3
½ lb. bacon ...	0	4½		19	6
¼ lb. currants ...	0	2			
4 lb. sugar ...	0	8	SUMMARY.	s.	d.
½ lb. tea ...	0	9	Rent	5	6
Yeast	0	1	Insurance ...	0	7
Milk	0	3½	Coal and coke ...	2	5½
1 lb. soap ...	0	3	House	0	6½
Salt and pepper	0	1	Food	7	5
3 oz. tobacco ...	0	9	Clothing ...	1	0
½ st. potatoes ...	0	3	Tobacco ...	0	9
Candles and			Debt	1	3
matches ...	0	1½		19	6

Third Week. Income £1 1s. Family Three.

	s.	d.		s.	d.
Rent	5	6	1 lb. bacon ...	0	9
Coals	2	2	½ lb. tea ...	0	9
Gas coke ...	0	3½	4 lb. sugar ...	0	8
Insurance ...	0	7	1 tin of milk ...	0	3½
Clothing ...	1	6	1 lb. soap ...	0	3
Meat	1	8	1 lb. starch ...	0	4
1 st. flour ...	1	5	3 oz. tobacco ...	0	9
¼ st. bread-meal	0	4½	1 box Globe		
Yeast	0	1	polish ...	0	1
1 lb. butter ...	1	2	1 box Zebra		
½ lb. lard ...	0	2½	polish ...	0	1

Third week—continued.

	s.	d.
Firewood ...	0	4
Blue	0	0½
Lamp oil ...	0	2
Matches and candles ...	0	2
Debt	1	0
Cat's-meat ...	0	1
½ st. potatoes ...	0	3
Bath-brick ...	0	1
£1	1	0½

SUMMARY.	s.	d.
Rent	5	6
Insurance ...	0	7
Coals and coke	2	5½
House	1	3½
Food	7	7½
Clothing ...	1	6
Tobacco ...	0	9
Debt	1	0
Firewood ...	0	4
£1	1	0½

Fourth Week.   Income £1 3s. 9d.   Family Three.

	s.	d.
Rent	5	6
Coals	2	2
Coke ...	0	3½
Insurance ...	0	7
Meat	1	6½
1 st. flour ...	1	5
¼ st. bread-meal	0	4½
1 lb. butter ...	1	2
1 tin of milk ...	0	3½
½ lb. lard ...	0	2½
1 lb. bacon ...	0	9
4 lb. sugar ...	0	8
½ lb. tea ...	0	9
Milk	0	1

	s.	d.
Eggs	0	2
1 packet gold-dust	0	1
1 lb. soap ...	0	3
3 oz. tobacco ...	0	9
½ st. potatoes ...	0	3
Candles and matches ...	0	1½
Lamp oil ...	0	2
Cat's-meat ...	0	1
1 box starch ...	0	2
1 pair boots ...	5	11
£1	3	9

Fourth week—continued.

SUMMARY.

	s.	d.			s.	d.
Rent	5	6	Food		7	8
Insurance ...	0	7	Clothing ...		5	11
Coals and coke ...	2	5½	Tobacco ...		0	9
House	0	10½			£1 3	9

Fifth Week. Income 19s. 6d. *Family Three.*

	s.	d.			s.	d.
Rent	5	6	Eggs		0	2
Coal	2	2	Milk		0	1½
Gas coke ...	0	3½	1 oz. wool ...		0	2½
Clothing ...	1	0	Carbosil ...		0	1
Meat	1	10	Stamp and blue		0	1½
1 st. flour ...	1	5	Debt		1	0
¼ st. bread-meal	0	4½			19	6
1 lb. butter ...	1	2				
1 lb. bacon ...	0	8	SUMMARY.			
½ lb. lard ...	0	2½			s.	d.
½ lb. tea ...	0	9	Rent		5	6
4 lb. sugar ...	0	8	Insurance ...		0	7
1 lb. soap ...	0	3	Coals and coke ...		2	5½
1 tin of milk ...	0	3½	House		1	0½
Yeast	0	1	Food		7	9
3 oz. tobacco ...	0	9	Clothing ...		1	0
Candles and matches ...	0	1½	Tobacco ...		0	9
Lamp oil ...	0	2	Debt		1	0
1 packet gold-dust	0	1			19	6

Sixth Week. Income 19s. 6d. Family Three.

	s.	d.
Rent	5	6
Coals	2	2
Gas coke ...	0	3½
Insurance ...	0	7
Clothing ...	1	6
Meat	1	8
1 st. flour ...	1	5
¼ st. bread-meal	0	4½
1 lb. bacon ...	0	8
4 lb. sugar ...	0	8
½ lb. tea ...	0	9
1 tin of milk ...	0	3½
Eggs	0	4
Yeast	0	1
½ lb. lard ...	0	2½
1 pint milk ...	0	2
1 lb. soap ...	0	3
3 oz. tobacco ...	0	9
1 box Globe polish... ...	0	1
1 box Zebra paste	0	1

	s.	d.
Lamp oil ...	0	2
Cotton and needles ...	0	1½
Candles and matches ...	0	1½
Potatoes ...	0	3
Debt	1	0
	19	6

SUMMARY.	s.	d.
Rent	5	6
Insurance ...	0	7
Coals and coke...	2	5½
House	0	10
Food	6	10½
Clothing ...	1	6
Tobacco ...	0	9
Debt	1	0
	19	6

The woman who drew up these budgets is a skilful, competent manager, and it is worth while for us to discuss her expenditure in detail, and see what her purchases represent. And as we do so we see that every penny has been considered, and laid out with wisdom and foresight, if we except, perhaps, one item, the 9d. for tobacco weekly for the husband, as compared with the 3d. which constitutes the weekly expenditure for milk. But this is the only item which can be called self-indulgent ; there is absolutely nothing allowed in this household for any form of diversion.

Opinions differ in these days as to the best form of diet : but presumably people of all shades of opinion will agree that, whatever the food selected, it is desirable to eat enough to satisfy hunger and to maintain strength. It is hardly to be expected that a woman in the position that I am describing should have a thorough knowledge of food values, or, indeed, if she had that knowledge, any wide possibility of choice ; but, on the whole, the diet is not unskilfully chosen considering the small funds available. In describing the quantities I give size as well as weight, as representing a clearer picture to the mind of the general reader. She buys, it will be seen, in the first week, in which her income is 18s. 6d., 1s. 6d. worth of meat. The sort of meat she would buy is 6d. or 7d. per pound. She has, therefore, about 3 pounds—that is, about such a quantity as would last over two meals for a family of the same size who were better off. One stone of flour makes, roughly speaking, 24 pounds of bread—that is, six quartern loaves, weighing 4 pounds each. If some of the flour is used for tea-cakes, it means correspondingly less bread. The $\frac{1}{4}$ stone of bread-meal represents 6 pounds more bread, that is, a quartern loaf and a half. The butter is what is known as Danish butter. It is cut in rough wedges. A wedge weighing $\frac{1}{2}$ pound is about 5 inches long by $3\frac{1}{2}$, and from $2\frac{1}{2}$ inches to $\frac{1}{2}$ inch deep. Two of these pieces, therefore, are the week's allowance. The $\frac{1}{2}$ pound of lard, which goes to make puddings, tea-cakes, etc., is about the same as the $\frac{1}{2}$ pound of butter. The week's allowance of bacon—1 pound—is a piece 8 inches long by $3\frac{1}{2}$ wide, and $2\frac{1}{2}$ inches deep —that is, the length and thickness of an octavo volume, but a third less wide. The 4 pounds of sugar, probably moist or granulated, represent ten teacups piled up with sugar. The $\frac{1}{2}$ pound of tea represents two breakfast-cups quite full of tea-leaves. This, compared with the rest of the items, is a liberal allowance, since $\frac{1}{4}$ pound per person is the quantity of tea usually allowed for a week in well-to-do households, and it is perhaps a pity that this should be the item, generally of an inferior quality, of which the supply is plentiful. The quantity of milk is small, 1 quart for the week, that is, enough to fill four tumblers, and including, apparently, cooking, bread-making, and every other purpose for which milk is used. The remainder of the food consists of vegetables: half a stone of potatoes, which will fill a packet 16 inches high and half as broad, and enough onions for flavouring.

Therefore, if we recapitulate the allowance of food

for the week for these three persons, as it would appear to the eye, we should see on the table—

> One dish of meat, tolerably full, not piled up, the dish about 10 inches long.
> Seven and a half quartern loaves, about 15 inches long, 8 in height, and 6 wide.
> Two wedge-shaped pieces of butter 5 inches long.
> One piece of lard of the same size.
> A piece of bacon about half the size lengthwise of a large octavo volume.
> Ten teacups full of sugar.
> Two breakfast-cups full of tea-leaves.
> Four tumblers of milk.
> A bag of potatoes 16 inches high and half as broad.

This is about the quantity, if we except the tea and sugar, that would last about two days for a family of the same size in better circumstances.

It will be seen by other items in the list that the woman, with the instinct of the housewife, spends on keeping the house bright and clean money that others might give to food. The first week we see 'Globe polish,' which is a sort of greasy paste used to clean brasses. This is a really more expensive item than it sounds, as the pennyworth is a very small round tin $\frac{1}{2}$ inch high. The packet of 'gold-dust' at first sounds a surprising item, but the gold-dust is a sort of yellow washing powder, looking like shredded soap. A packet of this, by the way, offered to a

housekeeper of another kind, was indignantly re-
pudiated as 'taking the skin off' her hands ; but the
writer of the budget described it as ' beautiful for
washing.' Matches, representing one box, and lamp
oil seem to be adequate. The amount of coal,
2s. 4d., represents two sacks, each about 3 feet high
and 1 foot across, and containing 1 hundredweight,
and this quantity also seems adequate.

It is worth while going on to consider the following weeks of Mrs. A. B.'s budget. The second week
she has 1s. extra ; that shilling goes to pay off the
debt, to which purpose 1s. 3d. is now contributed,
instead of the 3d. of the week before. The item of
clothing is still 1s. Half a pound of currants are now
added, and only half the quantity of bacon bought.

This week there is also 3½d. for coke, and 2d. less
for coals. The third week the income is £1 1s., and
the extra nearly all goes to household activity. Besides the pennyworth of Globe polish, we now have
a pennyworth of Zebra polish, which is a black
paste for cleaning the grate ; one pennyworth of
bath-brick for cleaning the step, etc.; one pennyworth of cat's-meat, and 1 pound of starch. The
fourth week the income is £1 3s. 9d. This week a
pair of boots at 5s. 11d. was bought—a heavy item
—and none of the debt paid off. The rest of the
expenses remain as before.

The fifth and sixth week the income drops again
to 19s. 6d. The expenditure varies a little, but

every new item added means something else left out,
as will be seen by the following table:

MRS. A. B.'s BUDGET FOR FIFTH AND SIXTH WEEK,
COMPARED WITH HER BUDGET FOR SECOND
WEEK. INCOME BOTH WEEKS 19S. 6D.

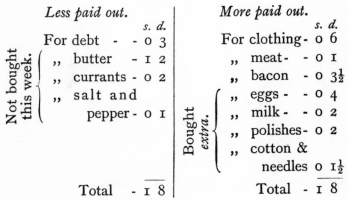

Fifth Week.

Less paid out.

		s.	d.
For debt	- -	0	3
,, insurance		0	7
,, currants	-	0	2
,, salt and			
pepper	-	0	1
,, potatoes	-	0	3
Total	-	**1**	**4**

Not bought this week.

More paid out.

		s.	d.
For meat	- -	0	3
,, bacon	-	0	3½
,, gold-dust		0	1
,, eggs	- -	0	2
,, new milk		0	1½
,, wool	- -	0	2½
,, carbosil	-	0	1
,, stamp	-	0	1
,, blue	- -	0	0½
Total	-	**1**	**4**

Bought extra.

Sixth Week.

Less paid out.

		s.	d.
For debt	- -	0	3
,, butter	-	1	2
,, currants	-	0	2
,, salt and			
pepper	-	0	1
Total	-	**1**	**8**

Not bought this week.

More paid out.

		s.	d.
For clothing	-	0	6
,, meat	- -	0	1
,, bacon	-	0	3½
,, eggs	- -	0	4
,, milk	- -	0	2
,, polishes	-	0	2
,, cotton &			
needles		0	1½
Total	-	**1**	**8**

Bought extra.

Mr. Rowntree, in his valuable book on York (p. 105), has said that it is possible for a man to live on 3s. 3d. a week for food, a woman on 2s. 9d., a child from eight to sixteen years of age on 2s. 7d., and a child from one year to eight on 2s. 1d. He has stated in the chapter on the budgets of the working classes that in the York workhouse, where the diet is regulated by the general order of the Local Government Board, the cost of feeding a family, consisting of man, wife, and three children, works out at almost exactly 6½d. per head per day. He adds : ' We must remember that the workhouse diet requires not only more cooking utensils, but also more trouble in cooking than the labourer's diet . . . that a large number of persons can be fed more economically than a single family, and that the cost of the workhouse diet is calculated at contract and not at retail prices '; and even so 6½d. per day means 3s. 9½d. per week, and 3s. 9½d. per week for three persons would mean 11s. 4½d. It will be seen, therefore, what a very modest allowance is the 7s. 4d. spent in food for one week by the family we have given above.

Mr. Charles Booth, in his great work on ' Life and Labour in London ' (first book of the first series, ' Poverty,' p. 133), gives an expenditure of 2s. 4½d. per male adult per week for food as the very lowest, poorest class. The budget of Mrs. A. B. goes but a fraction above this amount in expenditure, and yet the house looks absolutely clean and bright, and the man, woman, and child well turned out, which, considering the very small amount shown as expended on clothes, is a quite amazing result. It is very difficult to find out with anything approaching to accuracy the amount spent upon clothes. The six weeks of these budgets presumably were recorded at a moment when the family were tolerably provided with clothes, which needed but to be supplemented by the small amount stated. In the above list there is absolutely no margin for any extra expense of any kind, and in case of illness it means, as has been stated, absolute disaster.

This household can hardly be taken as a type; it is but an instance of what can be done upon the modest amount available. And even these, with all their care, are often in straits, at which times they get into debt, repaying it, as we have seen, by small weekly instalments.

The rent—5s. 6d.—always seems a considerable item for such an income. Out of 700 houses, 30 were under 3s. 6d. a week, 180 from 3s. 6d. to 4s., 254 between 4s. and 5s., 76 between 5s. and 6s., 62 between 6s. and 7s. 6d., 21 over 7s. 6d., and 3 men had bought their houses and were living in them.

The difference in the rent of these houses is a difference in position as well as of the quality of the house. There are some quarters which are more crowded and more insanitary, that are on lower land near the river, and in case of bad weather are flooded; these have lower rents. The houses at 7s. 6d. are nearly all in the newer quarters of the town, built in wide, airier roads, and with more accommodation. The houses mostly consist of two rooms on the ground-floor, two above, and a sort of little scullery or washing-place. The rent in nearly all the budgets I have been able to verify is entirely out of proportion to the rest of the expenditure.

One of the first things done when a workman's circumstances improve is to move into a better house. There are always expenses connected with the move; perhaps the good time may not go on, and then if there are other expenses, such as illness, and the man falls out of work, there has to be another move, and the result is disastrous.

It is somewhat difficult to obtain much information about the important item of expenditure on clothing, for in these cases also the happy ones are those who have least history. The careful manager, the skilful needlewoman, turns and re-turns and adapts; the unskilful and extravagant one sometimes pays sums that are almost incredibly out of proportion to the income. Thus it was found that one woman in receipt of 32s. a week paid 2s. 6d. a yard for the flannel out of which her husband's shirts were made (for these shirts it is absolutely no good buying inferior flannel), and then gave them out to be made by somebody else, and had to pay, of course, for the making as well. It is small wonder that where the clothing question is so solved the family budget should not be satisfactory.

In the house of the R.'s the man and his wife, from both being in ill-health, had got terribly into debt. Some friends put them straight, and offered to send the wife away for a change of air for a fortnight in order that she might be set up again, upon which she instantly got into debt again to the amount of £2 2s. in order to have good clothes to go away in.

The question of boots, both for children and grown-up people, is with the badly-off a constant difficulty, and one of the most serious that they have to face; and the miserable foot-gear of the women and children especially—the men are obliged to have more or less good boots to go to work in—is a constant source of discomfort and of injury to health. One reason why so many of the poor women go about with skirts which drag about in the mud is that they do not want to display what they have on their feet by holding their skirts up. A working-girl said on one occasion that she thought the mark of a 'real lady' was that she wore a short skirt and neat boots, this last representing to the working-girl almost the unattainable. Boiled-boot shops are still met with. 'Boiled' boots are old boots begged, found in the street, etc., picked up, patched, polished, and sold at a low price. There are various old-clothes shops, market stalls, hawkers' barrows, at which men's suits as well as women's clothes can be bought for a trifling sum.

The difficulty of paying for anything for which more than a very small sum in ready-money is needed explains the eagerness with which the house-wives of this town embrace any system by which they are enabled to buy in small instalments. Most of the women buy their clothes ready made, and pay for them and for their boots on the £1 ticket system. I do not know whether this obtains in other parts of the country. These £1 tickets are sold by men who buy them for cash down at certain shops in the town, getting the tickets for 18s. or even less; and the women, who buy these from them in their turn, pay 21s., payable in instalments of not less than 1s. weekly, and usually 2s. 6d. for the first week. These tickets are available either for one shop or two, sometimes 10s. goes to a boot-shop and 10s. to a draper. The advantage of this system over that of buying from the 'tallymen,' or hawkers, is that, although in each case the woman has to make a weekly payment, in the case of the £1 tickets she goes to the shop in the town and can get the goods that she sees at the prices marked in the windows, whereas by the other system she is at the mercy of the tallyman, who may palm off on her at a given price something which is usually sold far below it. She has, besides, to buy the thing unseen from a sample shown her. The question of good clothing at reasonable prices for the working-man is, in this country, a question which certainly appears to need a good deal of reform. The work itself of the people we are describing is, of course, ruinous to almost any sort of clothing, and it is not surprising that the man as he appears going to his work should wear the very worst clothes he can find. The man who, when his day's work is over, is going to turn himself out like an entirely respectable citizen is, as he goes off to his work in the morning, clad almost in rags, as, unwashed, unshaven, his can of food swinging in his hand, he strides along in a greasy, torn old coat with holes in it, patched trousers, frayed at the edge, tied tightly below the knee. So much for the outside garments; but it is essential that beneath these there should be something warm, and a warm flannel shirt of good quality is absolutely necessary. No cotton clothing is of any good, not only because the men often come away wringing wet, and are liable to be chilled unless they have woollen clothes on, but they have a theory that the gas and fumes 'eat' the cotton materials; indeed, when working trousers are bought, they are often turned inside-out and the seams sewn with wool before they are worn at the works.

One of the workmen told a visitor that he bought

a so-called flannel shirt, ready made, for 2s. 11d., and that by the time he had been in it three whole days at the works, where it was exposed to rain, to intense heat, to poisonous vapours, it simply went to pieces when he came back, and he described it as having 'flown up as fluff into the air' when he blew it. This man now buys shirts from 6s. 6d. to 9s. 6d., for he finds that nothing else is good enough. If he can procure an army shirt at a pawnshop he does so, and pays from 2s. 6d. to 3s. 9d. for it. A couple of these shirts will last a year. It is, at any rate, cheering to know that these are of such a quality that they are, on the whole, the most satisfactory things that the workman can buy.

I give here the budget of another family, the C.'s —one week taken haphazard among others of the family of a workman in receipt of 26s. a week, and with five children. Again here the rent seems excessive for a house of four rooms. The table given includes a balance of ½d. at the end of the week, happily on the right side.

C.'s BUDGET.

Family Seven (including Parents). Income £1 6s. per Week.

	s.	d.		s.	d.
Rent	5	0	Tea, sugar, etc.	2	5
Sick club and			Rice, peas, and		
insurance ...	1	6	barley	0	6
Coals, light, fire			Butter, lard, and		
wood	3	0	bacon ...	2	6

C.'s Budget—continued.

	s.	d.	SUMMARY.		s.	d.
Flour	2	10				
Butcher's meat	3	6	Food		12	11
Soap, soda, and			Rent		5	0
blue	0	6½	Insurance ...		1	6
Milk	1	2	Coals and firing		3	0
Boots and cloth-			Clothing ...		2	6
ing	2	6	House		0	6½
Sundries ...	0	6	Sundries ...		0	6
	£1	5	11½		£1	5	11½
Balance to			Balance to			
credit	0	0	0½	credit	0	0	0½
	£1	6	0		£1	6	0

Here is another budget of a family of six persons —parents and four children. There is, as will be seen, a balance of 2s. 4d. unaccounted for at the end of the week.

D.'S BUDGET.

Family Six (including Parents). Income £1 4s. per Week.

	s.	d.	SUMMARY.		s.	d.
Rent	4	6				
Provisions ...	9	6	Rent		4	6
Butcher's meat ...	4	0	Club		1	2
Coals and light ...	2	6	Food		13	6
Club	1	2	Coals and light		2	6
Unaccounted for	2	4	Unaccounted for		2	4
	£1	4	0		£1	4	0

Here is the budget of a family in receipt of £1 10s. a week. This family has four children, and we have a total of 16s. 2½d. per week for food for two adults and four children. This man, as will be seen, not only pays 9d. towards insurance, but puts into two sick clubs besides.

E.'s BUDGET.

Family Six (including Parents). Income £1 10s. per Week.

	s.	d.
Rent	5	6
Club and insurance ...	2	6
Coals and light	2	9
Clothing and boots	3	0
School	0	6
Meat	3	6
Groceries ...	6	0
Flour	3	6
Potatoes ...	0	7½
Milk	0	7
Eggs	0	6
Bacon	1	6
£1	10	5½

SUMMARY.

	s.	d.
Rent	5	6
Club and insurance ...	2	6
Coals and light	2	9
Clothing and boots ...	3	0
School... ...	0	6
Food	16	2½
£1	10	5½

Spent in excess of income, 5½d.

The following family have £1 1s. 9d. per week for father, mother, and one child.

F.'s Budget.

Family Three. Income £1 1s. 9d. per Week.

	s.	d.
Rent	5	0
Coal	1	8
Ferry tickets ...	0	6½
Pocket-money	1	2½
Groceries ...	10	0
Dog-biscuits ...	0	2½
Meat	0	9
Fish	0	3
Milk	0	7
Cakes	0	4½
Sundries ...	1	2
£1	1	9

SUMMARY.

	s.	d.
Rent	5	0
Coal	1	6
Pocket-money	1	2½
Ferry-tickets ...	0	6½
Dog-biscuits ...	0	2½
Sundries ...	1	2
Food	11	11½
£1	1	9

G.'s Budget.

Family Three. Income 19s. 6d. per Week.

	s.	d.
Rent	5	9
Coals	0	4
Doctor	1	0
Trams	1	0
Ferry-boat	1	6½
Left for food	9	10½
	19	6

This leaves a total of 9s. 10½d. out of the wages for the man, woman, and boy of sixteen to live upon.

The families whose budgets have been given above are, as will be seen, very near the margin of the absolutely poor.

All the above tables, it is evident, represent the thrifty part of the population, those who may be generally classed among the deserving; but I fear it would be idle to pretend that this description can be applied to most of the homes at the works. The all-devouring tendencies to drink, betting, and gambling, the main channels of the waste and leakage of the workmen's funds, will be discussed in another chapter.

Then there is a form of expenditure frequently met with which, if it may not be wise, is generous and beautiful—the amount expended on charity by the very poor, who, with self-sacrificing kindness, seem constantly ready to help one another. It often happens that if one of their number is struck down by accident or sudden illness, a 'gathering' is made at the works, the hat is passed round, and each one contributes what he can to tide over the time of illness, or, in case of death, to contribute towards funeral expenses.

The expenditure on funerals—cabs, mourning, etc. —is usually greater after an accident, as it appeals to the public imagination more. Even when a man has been insured, and there should therefore be a small sum to tide over the first moment of great need, it often happens that nearly all the insurance money goes in the funeral. 'I put him away splendid,' you will hear a widow say, forgetting, or at any rate accepting, the fact that her house is nearly bare of necessaries, and that in a day or two she may not know where to turn for bread. Another said with pride after her husband died that she had 'buried him with ham,' meaning that the assembled company who came to the funeral had had sandwiches of the best description.

A funeral, indeed, is one of the principal social opportunities in the class we are describing. 'A slow walk and a cup of tea' it is sometimes called, and the busy preparations in the house for a day or two before, the baking, the cleaning, the turning-out, are undoubtedly often tinged with the excitement and anticipation of the entertainer. And after all we must not forget that to many women, at any rate, giving a party, having a great many people in the house at once, is in itself a stimulus and a pleasure, and that for those of the community who are debarred by their conditions as well as their habitations from giving an 'at-home' or a dance, the justifiable crowding of the funeral means absolutely the only opportunity for keeping open house, and is accordingly eagerly seized.

On one occasion a woman whose husband had been sent to a hospital in London went up to nurse him. When, shortly afterwards, he died, a few pounds were sent her to pay for the return journey and to tide her over the first weeks after his death. A few days after, a visitor who happened to be in the station on the day of her return was stupefied at seeing a figure alight clad in the mourning weeds of the stage, including a long black skirt, a deep crape flounce, and everything complete. The mourning, and what she called a ' proper ' funeral had absorbed nearly all the margin which should have kept her for a month, and in a few days she was again in straits, with a piteous request for more help. And the difficulty is that, since these wild outbursts of expenditure generally take place in a crisis of emotion, it is not easy at the time to preach against them.

From the women who, although well-intentioned, are not skilful managers, or from those whose expenditure is neither exemplary nor well-intentioned, it is obviously not so easy to obtain details of expenditure. It is only possible to guess whether the wages are sufficient by having means other than direct inquiry of knowing the amount of income, by knowing the number in the family, and observing the general appearance of the house and its inmates. In over one-third of the houses visited the women did not even know what their husbands' wages were. The criterion of whether a man is a ' good husband ' or not is often, in public opinion, the proportion of his wages which he gives to his wife; and, indeed, it is a tolerably good rough-and-ready rule, since, although the husband may occasionally withhold the money because he believes that the wife would waste it if she knew the extent of his resources, she is more often kept in ignorance because the husband wishes to have the proportion he chooses to do what he likes with. In many cases where the husband makes over the whole of his wages to the wife, it is agreed between them that she shall return a fixed proportion to him for his own pocket-money and personal expenditure; and this, whatever else happens, he reserves for his own use, and is entitled in the opinion of both to spend without a qualm. In the budget of the A.B.'s, which stands at the head of this list, it will be seen that the husband has 9d. weekly for his tobacco. In the case of the D.'s, 2s. 4d. unaccounted for was probably kept back by the man without any statement as to what it was used for. I have seen one household of a man, H. R., his wife, child of three, and a baby, in which the man was in receipt of 42s. weekly. Of this the wife gave him back 7s. to ' waste '—it was so explicitly stated—and on further inquiry the ' wasting ' consisted in spending it on sweets, on the theatre, or on the music-hall. This is a most respectable, well-conducted man, who spends his Sundays in bed and reads all day long, neither smokes nor drinks, and probably, therefore, feels justified in spending 1s. a day for his diversion.

S. D., whose wages amounted to probably between £2 and £3 a week, was in the habit of putting 30s.

into one pocket always for his wife. He said: 'Whatever happens, I put 30s. in here for her, and she needn't mind what is in the other one.' This arrangement appeared to satisfy them both. B. M., who entirely declined to tell his wife how much he earned (this man, I believe, had nearly £3 per week), always gave her 30s. He also put 5s. a week into the savings bank, and subscribed 9d. per week to two sick clubs, but he made his wife, out of the 30s., pay his insurance policy, saying, 'That is her business; that won't benefit me, so she must keep it up.'

F. G., who has 45s. a week, gives it all to his wife. She allows him 1s. a day pocket-money, which he spends on sweets and chocolates. She also insists upon his paying 3d. a day for his stout, which she considers 'a luxury and quite unnecessary.' Another man (these are young married people who have one child of five months old) earns 50s. to 68s. a week. He gives his wife 28s. to keep house on, out of which she pays his sick club. She appears to be a careful and thrifty woman, and to do very well on it. What is done with the balance of her husband's wages is not stated. Another man, a labourer, who has 25s. a week, gives it all to his wife, and she gives him 2s. a week for pocket-money, but he rarely spends it all. Instances of this kind might be multiplied, and, as will be seen, where the woman has the upper hand, in spite of the wages being earned by the husband and her receiving them from him, she makes a favour of the amount she gives him back again. One man, who sometimes has an extra shift, or even two, in the week, brings home that extra money separately wrapped up in paper, and he and his wife make a point of keeping it entirely to spend upon the children, of whom he has seven; it never goes for the rent or for any necessaries for the father and mother. This man has an average wage of 45s. a week.

The plan of setting a definite sum apart in the way we have described to pay for diversion and change of thought is probably less prejudicial to the slender purse than the unavowed opportunities for spending, the constant temptation which comes in the way every day, every hour, of each man and woman in the works, even if this temptation may be considered spending of a desirable kind. The tallymen hawking up and down the streets come round to the doors of the workmen's houses offering all kinds of wares on the hire system, wares of a motley and amazing description: a mangle, a thimble, an umbrella, a china dog, a writing-case, a cupboard, a piano, a gramaphone—I have known of all these things being bought, and the variety of purchases on this plan is infinite. The part of the world we are considering is very musical, it is true: but in many cases the musical instrument has been bought simply as an adornment to the house. A visitor, hearing that Mr. and Mrs. M. did not care about music,

asked them why they had a piano in their house, to
which they replied, ' Oh, that isn't music, it's furniture.' And yet those who deny the advisability of
the workman's spending on anything but bare
necessaries, who deny his right to do so, may be in
danger of taking from him and from his home the
resources which will help to keep him there. The
more he adorns his home the better—if he has the
money to do it. That is, no doubt, a big ' if ': but
still it is hard on him that economics should label as
wrong in the artisan a course of action which to the
prosperous is a constant and legitimate source of
delight. But in many cases it happens that the
piano or the harmonium is not only furniture, but
music as well.

In one house, where the father, a blacksmith, was
very musical, as were also his children, the eldest
girl had a sweet, true, soprano voice, and when he
came in from his work he used to teach her to sing,
accompanying her songs on the harmonium. In
another cottage, on a winter's afternoon, a visitor
found a poor woman quite alone in the twilight
crooning a hymn, which she was endeavouring to
pick out at the same time on an old concertina, and
in reply to the visitor's comment, she said, ' I've no
little uns, you see, and I must put over the time.'

It is with the women that the tallymen, who often
come round when the men are away at work, drive
their best trade. If a woman is in need of ready-

money she buys on the aforesaid hire system, by which the purchaser pays so much a week until the whole sum is paid off, something that she does not need in order that she may pawn it. This gives her at the moment a small lump sum. But if she misses one week's payment, both the purchase and the weekly deposits are forfeited. And the situation is still further complicated by the purchase having been pawned and her being unable to redeem it; for, after all, pawning, which is often branded as a sort of crime, is to many of the people we are describing the one way of getting a little extra ready-money, which they need just as often for legitimate purposes as to pay a gambling debt. If any extra nourishment, for instance, is ordered for the sick child whose father is living, or trying to live, on 20s. a week or a little over, the only way to get that nourishment is to pawn something. I have seen people who have pawned their underclothing in order to keep up a respectable appearance outside. One woman habitually pawned her husband's clothes every Monday, received 3s. 6d. for them, and took them out on the following Friday, paying 4s.

In one house a little child lay dead, and the parents, unable to pay for its burial, finally pawned their clock that was hanging on the wall, which was about the only thing available left in their bare little house. There is absolutely no credit given for a funeral; for that purpose ready-money must somehow be found. It is a heartrending and constantly repeated experience to see in a workman's home where there is sickness, especially if it be the bread-winner himself who is ill, the house being gradually stripped, as, with each fresh need for ready-money, one thing after another is taken to the pawnshop.

CHAPTER IV

ILLNESS AND ACCIDENTS

THE question of health, so important a factor in the existence of all of us, assumes still greater importance in the lives of the workmen, to whom even a passing ailment means either a diminution of the weekly income, or else a continuance of work under conditions which may turn the slight indisposition into something more serious. The advice to 'take care,' that vague medical admonition that is so often a euphemism for a definite verdict of a more discouraging kind, is here but a mockery, for with most of the workmen such care is impossible. The well-to-do man may wake up one morning, it may be, not feeling as well as usual, simply remain in bed, take his own temperature, perhaps, and telephone for his doctor to come and see him. He is then advised to stay at home and have a day or two's rest, warmth, and quiet, at the end of which he is probably well. But the workman has no telephone, and needless to say no thermometer; he subscribes, it is true, to the doctor belonging to the works, whose services he is entitled to use, but he probably does not know where to lay his hand on that doctor, and in many cases his idea of finding him is to wait until he casually meets him in the street. He goes out, therefore, as usual, no matter what the weather or how he is feeling, in order not to lose a day's work and a day's pay, and when he comes home it is not surprising if he is worse. Then, finally, the doctor is perforce called in, and his verdict is accepted in a helpless, bewildered manner, often without understanding it, without having information, experience, or assurance enough to question it or suggest anything else. And then follows an illness in which mental suffering is bound to be added to all the rest: a time in which physical discomfort and wretchedness, the inconvenience of having daily life interrupted, so keenly felt and complained of by the man who can afford to be idle, are intensified tenfold in the case of the workman by anxiety at his pay being stopped at a moment when he needs it most. A foreman, if he falls ill, receives his wages all the same for the first two weeks he is off work ; the ordinary workman does not. For even if he is in a sick club his income is lessened at a moment when it should be increased, and the food and remedies that are desirable are in many cases unattainable without sacrificing something essential to the welfare of the rest of the household. In many cases the surroundings of the sick man are

such as to retard recovery.   There are of course
found among the workmen's wives, as in other
classes, women who are born and skilful nurses:
but many a time they are helpless and incompetent;
the house in which the patient is lying is often
crowded, noisy, stuffy, and dirty, and more so still in
times of illness than at others.   For illness brings
more for the housewife to do, unaccustomed duties,
more trouble: the routine of the house, such as it is,
is broken into, everything is bound to be even more
uncomfortable than it was before.

At a house in B. Street, where the father and
three of the sons were all ironworkers and had
worked for the same employers for many years, one
of the sons, a young man of twenty, was ill with an
incurable disease.   His bed, to avoid so much going
up and downstairs for the people who waited on him,
had been brought down to the kitchen.   The family
consisted of the father, mother, and eleven children,
of whom the invalid was the second.   The whole
family were living at home in a four-roomed house;
some of them slept in the same kitchen where the
boy was lying, all of them dressed there, ate there,
talked and laughed, squabbled and fought there; the
cooking was done in the same room; some of the
household washing had been done and was also
being dried there, for it was a rainy day outside and
there was nowhere else.   In this case the condition
of the invalid was in any case hopeless.   It was dis-

cussed quite freely in that sense before him, indeed,
his family saying that they were going eventually
to move into another house in a healthier quarter,
but could not do so as long as he was alive. But if,
instead of his case being hopeless, he had been in the
state in which his recovery and well-being depended
on careful tending, quiet and favourable conditions,
what chance would he have had? Over and over
again one sees cases of this kind.

And when—or if—the man begins to get better,
new difficulties are encountered at the critical time
when convalescence is beginning and strength ought
to be kept up. The doctor prescribes light and
nourishing food—eggs, milk; it is almost impossible to obtain them. The process of recovering
strength is slow and wretched. The man goes back
to his work too soon, and then a very short time
afterwards, perhaps, has to knock off again. The
same thing happens when there has been no regular
illness, but the general health has from one cause
or another become affected; the workman is what
is called 'below par,' going about listless and dis-
heartened, dragging himself to his work as best he
may; and if it is somehow compassed that he should
go away for a change, he loses his wages while he
does so.

In the case of any complaint necessitating an
operation, the workmen go into the very admirable
Infirmary in the town, leaving wife and children, if

they have no other resources, to struggle with life as best they may. When one considers what a sudden illness means to the well-to-do, and especially all that being told of the necessity for an operation means—the sudden revelation of unexpected danger, the swift disintegrating of existence, the way that everything falls into another focus—it is appalling to realize what this must mean in a case where illness and the sudden dread anxiety are the least part of the misfortune.

I remember one home in which from one day to the other the worker had suddenly become aware of having a complaint in his eyes which might necessitate an operation. He had been off work for some time. Half blind, hardly realizing what was happening to him and what would happen next, he was almost distracted with agitation, the poor little wife also in a state of dazed bewilderment, trying to do what she could, and hardly conscious of what she was doing. These people had had absolutely no visitors from outside; with this appalling news descending on them, they were struggling dumbly alone to endure it, and wondering what was going to happen next.

Another man was threatened with cataract in one of his eyes. The idea evidently uppermost in his mind was that the fact should be concealed, in case some one should insist upon his being treated; for his wife was on the verge of being confined, and he could not afford to be off work until the event was over.

Many of the working people, instead of going to the doctor, consult the 'herbalist,' in whom they have great faith. His shop is full of remedies. He sells them a book in which is a list of ingredients. He then sells them certain of the ingredients, either recommended by himself or chosen by them from the list, and they make the decoction themselves. It is, after all, a debased and modern form of gathering simples, and occasionally, when the required treatment is but elementary, when it is a question of a soothing potion for a cough or a fomentation, it is found to be efficacious.

One is apt to be surprised at first, considering that it is presumably the strong and stalwart who have taken up this work, to find how many of the workmen are more or less ailing in different ways; but we cease to be surprised when we realize how apt the conditions are to tell upon the health even of the strongest, and how many of the men engaged in it are spent by the time they are fifty. To say that this happens to half of them is probably a favourable estimate. They are exposed to extremes of temperature, being liable to become violently heated when at their work, and violently chilled when they move away from it. They come home tired, their vitality lowered, their clothes often wringing wet. They are constantly inhaling noxious fumes. The men working near the furnaces, especially those working near the level of the ground where poisonous gases are hanging about, are very liable to be affected by them, the man becoming drowsy, and eventually, in severe cases, unconscious. But it does not generally go so far as this; any man working near the furnace knows at once when he is suffering from the effects of the gas, and takes steps to get into the fresh air. An attack of this sort causes real discomfort, as even in a slight attack it gives a feeling of sickness and a bad headache. Sometimes, also, the ' chargers ' at the top of the furnace are ' gassed,' but only to a slight extent.

The men also suffer from rheumatism, from asthma, from pneumonia (often of a dangerous and virulent kind), from feverish attacks, from blood-poisoning in one form or another caused by some scratch on the surface of the skin when handling hot iron, from affections of the eyes due to exposure to dust, to glare, and to noxious vapours. Consumption is also frequent. It happens over and over again that, when it has been possible to arrest the disease by sending the man away to some sanatorium, he has fallen back as soon as he has returned to his unhealthy surroundings. The saying that has so long been an article of faith with us all, that every man is either a fool or a doctor by the time he is forty, is not true of the workman; or at any rate, even if he has been able, arrived at that stage, to frame for himself some sort of a healthy and desirable rule of life, he cannot well carry it out. The conditions of his life are in direct contradiction to those elementary rules of hygiene by which the well-to-do, as a matter of course, are accustomed to govern theirs. As to diet and digestion, indeed, let us hope that the concentrated attention on them which is possible and congenial in these days to so many of the prosperous is not absolutely an essential to well-being, for in that case the workmen run but a poor chance. Many of their wives practically do not know how to cook at all, or at any rate do not wish to do so, and get their food ready cooked from an eating-house or a fried-fish shop; and some of the others who do cook in the home do it so badly that it might be almost better if they did not attempt it. On the other hand, there are, happily, many of the women who can cook fairly well, and, according to their lights, take a great deal of trouble about it. The question is complicated by having to arrange not only for the family meals in the house, but for the food the husband is to take over to the works with him. The result of the food having to be carried over is that it is often of a sketchy and rather unsatisfactory kind, not, perhaps, in the households where the wife is competent, but in the larger number where she is not. The kind of housekeeping, if one may give it that name, that one too often comes across, the absolute incapacity or disinclination of the woman to bring anything like method or regularity into the family habits, is a revelation of domestic disorder.

In one house in S. Street the mother and two children, aged three years and nine months respectively, were having for breakfast cold pork and cheese. In another house, at 10.30 a.m., with absolutely no washing even of faces or preliminary toilet at all, and the children not dressed, they had just had for breakfast (after which the woman, leaving the dirty breakfast-things all about, was beginning to bake) cold ham, cheese, pancakes, marmalade, and bread-and-butter. All the cooking was done by gas, on the penny-in-the-slot system. One pennyworth of gas will cook two separate lots of steak or chops ; burned for light only in the room, it lasts eight hours. These people had 30s. to 35s. a week, and there were two young children. It seemed to be a happy home, albeit the woman was evidently not as skilful in management as many of the workmen's wives, since on this sum she found a difficulty in making both ends meet. This might be explicable if the other meals were on the same somewhat magnificent scale as the breakfast.

Another family, visited at 10 a.m., were just going to have breakfast, consisting of tea, home-made scones, pickled herrings, and butter.

The morning meal of the workman who is on the 6 a.m. shift, prepared overnight and left for him to take in the morning, is nearly always tea, ready mixed, and cheese. In one or two of the houses the tea had actually stood during the night, but it is usually poured out, mixed with milk and sugar, and covered up. In nearly all these houses condensed milk is used; it goes a great deal further than fresh milk, for a threepenny tin of condensed milk will last a fortnight.

The man who is working on the eight-hours shift, as described, and whose household resources enable him to have a solid meal, can count, as far as hours are concerned, on having it at home at one time or another, as if he works until two o'clock he can have it when he gets in in the afternoon; if he begins work at two he can have it before leaving. Besides this, he of course takes food with him to the works, as also do the twelve-hours men who do not work on the shifts. The diet varies, naturally, with different people. I take some instances haphazard.

One of the twelve-hours men, a man in A. Street, takes with him to the works for breakfast some bread-and-butter, some tea in a paper packet, and some sugar. This man does not care for the customary cold pies or cold meat for dinner, so he always has cake and hot tea for dinner in the middle of the day at the works; then, when he gets home in the evening, they all sit down together to a hot meal of meat and potatoes. The meat varies—pork, mutton, beef; but it is always hot. It is bought and cooked in small quantities just for the meal. For supper, about nine o'clock, cheese, cake, or 'something that is not meat or fish.' This man did not 'believe in much meat.' A man in C. Street does not consider his mid-day dinner (taken over to him at the works by one of his children) complete unless he has meat and potatoes as well. 'It's no dinner without there's potatoes.'

A woman in F. Street seemed to be extremely pre-occupied by the necessary cooking. She said she has no sooner got the tea cleared away and the things washed up than it is bedtime, then the husband wants supper, and the next morning's food has to be prepared to be taken with him, if he is on the night shift, at 10 p.m., or the next morning at 6 a.m., if he is on the morning shift. He takes meat or cooked bacon or ham and eggs with him. This woman bakes at home, as many of them do.

Another woman was just preparing some fried sausages for her son's dinner. This seems to be a favourite dish.

In a house in G. Street food seemed to be the most important factor and chief topic of conversa-tion. The woman said that, in the hot weather especially, she had to do a great deal of 'brain-thinking' as to what was best to put up for her husband's meal. His job at the works keeps him from 6 a.m. to 5 p.m. every day. He is not on the eight-hours shift. The wife gives him for the mid-day dinner cold meat—pork, beef, mutton—or cold

pies : the latter can be heated, if preferred ; also for
this meal cold tea already mixed with sugar and milk.
But for the other meals at the works he takes tea and
sugar wrapped up in paper and does without milk.
Sometimes he has eggs instead of meat, as he is a
Roman Catholic. Sometimes on the other days he
does not 'fancy' meat, and then he has tomato sandwiches, also a favourite form of food with the men.
Sometimes, again, if he does not care for what she
sends, he eats nothing, and if on such an occasion he
has to work all through the night, which sometimes
happens, he goes without food till the morning, as it is
too far to go back for more supplies—not the best condition in which to go through strenuous physical
labour.

Pickled cabbages are a favourite form of food.
A man in E. Street said he never counted he had
had a meal unless he had had pickles with it, or ' to '
it, as they say in Yorkshire. The favourite kind are
big red cabbages, of which four large ones pickled
in a glass bottle would last the family one week, as
an accompaniment to the usual fare of bread-and-
butter, cheese, meat, sausages, bacon, pies, according
to the meal-time.

A man in K. Street, if at home for breakfast, has
bacon, eggs, sausages ; if for dinner, a hot meal with
potatoes, and a hot supper if going on the night
shift, which begins at 10 p.m. The woman in this
house was cooking fried steak for her own dinner and

the children's.   Their income was about 24s. a
week.

A man in M. Street, when on the early shift, takes
a sandwich of cold bacon and bread for his breakfast,
and one of bread and butter.   The family have bread
and butter for dinner.   The man has a meat dinner
(hot) at 2 p.m. ; and anything left over is either
heated up or given cold to the children at tea-time.
When the man is on the afternoon shift they all have
a meat dinner at twelve.   On Fridays he takes cheese
to the works and has fish at home, as they are Roman
Catholics.

Whatever the variations of diet, a large amount of
tea almost invariably forms a part of it.   This is
mostly Indian tea at 1s. to 1s. 4d. a pound, and it is
bought in small quantities, usually in packets of
about 1 ounce, sold for 1d.

The eight-hours men have a quarter of an hour
for each of their meals, and must have them when
and where they can.   The men who are working
straight through the day have half an hour for breakfast, and some half an hour, some an hour, for
dinner.   Each of the 'gangs'—that is, each of the
several divisions of labour — the boiler-cleaners,
furnace-keepers, weighmen, etc., have a 'cabin,' a
little shed to which they retire for shelter during the
time off.   Here a man leaves his jacket hanging up
when he goes to work, he sits and smokes his pipe
when he has a rest; and if there is a stove in the

cabin, he keeps his can of food warm on it. The whole thing, according to our national methods, is extraordinarily haphazard, and on reading such an account one's fancy turns at once to the possibility of starting some admirable central place where good and cheap food should be provided. But the question of catering for the men during their day's work is not as simple as it looks, given the varied hours and different lengths of time available for the various workers. This is the kind of matter on which it is not always easy in this country to meet the views of the workman. He may possibly be aggrieved if he is not arranged for and helped a little; but he will certainly be angry if he is arranged for and helped too much. One learns after long experience that what he wants is that the next step that is of any concern to him, and that he chooses to take in his own way, should be facilitated for him; but one has to be very careful not to go a little further and make superfluous suggestions, which he then, much to the discomfiture of the suggester, absolutely negatives and dismisses. The men at the works in question prefer, and naturally, not to have far to go for their dinner; and on the whole they like to take it with them, and sit down to eat it wherever they may happen to be. They do not wish to have while they are at work a room to eat in where spotless tables would seem too clean for them comfortably to rest their blackened arms upon. It is easy to understand

that the man who gets home from his work in such
a condition that before looking like a respectable
citizen he has to make an elaborate toilet, washing,
and changing all his clothes, does not care to do that
more than once in the day, and certainly does not
want to do it while he is at work.   On one occasion,
when it was proposed to start some sort of catering
establishment for the men, where they could either
keep their food hot or buy a well-cooked and cheap
meal, they were canvassed to know whether they
liked the idea of the scheme.   A little more than
half of them replied that they would like it, but most
of them added an emphatic rider to the effect that
it would be a good thing so long as it was 'handy
enough' to go to straight from their work without
a wash, and without wasting any of their precious
spare moments in transit; while the rest of them
resented the idea that their usual methods could be
improved upon by outside organization.

In the rolling mills of the steelworks, where the
ingot, or great block of steel, is rolled out, becoming
narrower and narrower, until the short, thick block
is turned into a long, thin, red-hot rail, the end of
that rail is cut off in order to make it trim and
square, and the men who are working in the rolling
mills use the pieces cut off as stands to keep their
dinner hot upon.   One of these who was addressed
on the subject of improving his dinner arrangements
entirely scouted the idea of any other provision being

made for his comfort than that which seemed to him entirely adequate of the little red-hot pedestal on which his can was placed while he worked, and beside which he sat down contentedly to eat when the moment came.

I am bound to say that this regimen, the kind of food usually taken over to the works and the conditions under which it is eaten, appears to be in its working and results as often successful as the reverse.

Besides all that we have shown in the condition and surroundings of the workmen as affecting health, we also have to consider the undeniable risk, the constant possibility, of accident with which all these men are practically face to face—a danger that the onlooker is apt, perhaps, to minimize by saying, ' He can avoid it if he takes care.' That is to say, the man to whom the accident happens has relaxed for a moment the incessant alert watchfulness, as necessary to the ironworker as to the sailor. But it is of no good saying that ' by care ' this or that accident can be avoided, or postulating as a condition something which human nature does not seem to find possible. All experience shows that the watchfulness, the vigilance, the alertness, is not—as a matter of fact, cannot be—incessantly kept up. The possibility of accident, the incessant confronting of danger, is a thing which becomes too familiar to the ironworker; and there comes a moment when, from fatigue, perhaps, from strain, from the immunity which may have attended the perilous job for a long space of time, attention is relaxed and a disaster happens. There are few streets, probably, at the works, or in the town adjoining, where some of the inhabitants have not been at any rate in proximity to one of those experiences, of which the recoil and the thrill are felt by all of them as a recurring constant possibility to themselves. Happily, serious accidents are comparatively rare; comparatively, that is, when one considers the incessant possibility of them. The mere weight of the material and tools with which the ironworkers work, the continual risk of injury from sparks, from explosions, from burning, means that any false step or stumbling may here have serious consequences. I take some instances haphazard over a period of three years to show the sort of thing that may happen. One man had had his legs scalded from hot steam escaping from a boiler; another had his finger crushed by a piece of iron falling upon it; eyes had been injured by explosions; arms had been amputated in consequence of accidents to them; one man had his foot crushed by a loaded barrow falling on it; one twisted his thigh by slipping while he was charging the furnace. A workman injured when employed in or about a works on his master's business is now entitled to be paid after the first week not less than 50 per cent. of the average earnings for the previous twelve months, up to £1 per week. If the incapacity to work lasts more than a fortnight, payment is to date from the day of accident. Three years ago one of the most terrible of all accidents occurred, in which two men fell into one of the blast-furnaces. The bell, the metal cone already described, on to which the charge is put before it is lowered into the furnace, was being changed, and the old bell was being lowered into the furnace. As it was lowered some accidental contact produced an explosion, and a man who was standing regulating its descent on a plank fastened to the side of the furnace fell off into the glowing mass beneath, where the temperature is about 3,000° F. Another who was with him on the plank, who had retained his footing, absolutely lost his head, helplessly stepped forward and met the same fate. A third, who was nearer the side of the furnace, was seized, and managed, half dazed, to scramble out, after living through the moment in which the same fate might have overtaken himself, although not through what must have been the wild and unimaginable agony of his fellows in that second before they became unconscious. Then followed two hours of horror for those who, standing at the mouth of the furnace, tried to recover the remains, and at last succeeded in doing so; two hours during which those engaged in the search and directing it needed the very utmost of their will and self-control not to give way in sheer nervous collapse. A clergyman had been sent for to read the Burial Service at the mouth of the furnace in case the remains had not been recovered. In mines and coal-pits, after such an accident, nothing more is done during the day, and the mine or pit is laid idle. But at the ironworks the work cannot be interrupted; the furnaces cannot be allowed to go out; the regular succession of all that is done cannot be stopped; and the men who have seen the catastrophe which has overwhelmed their comrade go back, sick at heart, to continue their own work. To most of the women who live in those crowded little streets of iron-workers the risks of that life are ever present and ever anticipated, and the simple heroism and endur-ance with which both anticipation and fulfilment are faced are a constant source of wondering ad-miration to those who look on. When the blow falls, it means that all the details and surroundings of life, the material possibility of being able to exist at all, have been struck at, as well as the centre of sentiment. After a death, either from accident or illness, the neighbours all instinctively draw together to show their sympathy and participation. If you go into a house visited by death you will find it full, quite full, of visitors. On such an occasion the dead body is lying in the room which is at once kitchen and general living-room of the cottage. All round the walls, on three sides of the room, wherever there is available space, people are seated, tightly

wedged together, sitting sometimes in silence, sometimes bringing out simple inarticulate sentences of
attempted consolation. On such occasions the room,
as much of it as can be draped, is hung with white,
also the bed and the table standing near it. On the
table sometimes stands a plate with a pile of new
tobacco-pipes on it; another, perhaps, with plum-
cake or biscuits; and in Roman Catholic houses tall
candles to light during the night. The men who
sit round will smoke in silence.

And when the funeral takes place all these neighbours will follow it, with every mark of sympathy and
respect. Many a time, when visiting a home thus
stricken, in those moments of forlorn reaction worse
to endure than the first deadening sorrow, one gets
from a few simple heartrending words the impression
of what the now desolate home has been before the
mainstay was taken from it—the beloved husband or
son who was the bread-winner as well. A widow
whose eldest son, her remaining support, had been
killed at twenty-two, said that whenever the men
came in late for a meal she always thought, according
to the current phrase, that they had 'happened an
accident.' This story was a pitiful one. The boy
had been killed by an engine turning over upon
him; he had taken a job that was not his own to
fill the place of another man who was off that day.
The lad was strong and capable, an excellent son,
the pride of his mother. She sat wringing her

hands, saying, 'He said to me only this morning, "Mother, I'll make a lady of you yet," and he meant to do it, he did.'   And a boy of ten, to whom the lad who was gone had evidently been a great hero, was standing crying forlornly—perhaps wondering whether he would be able in the time to come to do what his brother had meant to do.

Another poor woman, whose husband by a false step had fallen from a scaffolding and been killed on the spot, was sitting in blank misery in her house, and talking of what he had been to her in little disjointed sentences that painted a picture eloquent and complete.   'He was such a man for his children.   He thought all the world of them.   He'd always have his children round if he went anywhere.   Such a quiet man he was, all for his home.   He didn't hold with going out of it unless it was to the parish-room. Yes, he thought all the world of his children.'   And the eldest of these, a girl of fourteen, stood by listening with a face of hopeless misery.   The house, which had always been a pattern of neatness and comfort, looked strangely transformed; the wife, with no reason to have anything done at the appointed time, sat bewildered and rudderless, wondering what would happen to her next, for her home would probably be broken up as well as her life.

The question of the widows is always a difficult one.   The widow of an ironworker, like the widow

of a clergyman, or of a man of any other calling
whose abode depends upon his work, is perforce
turned out after her husband's death to seek for
another home, so that the house may again be
occupied by a workman, unless, that is, she has
sons who are working and justify her remaining.

One man lost both his eyes in an explosion. He
escaped with his life, if that meant anything after
the terrible thing that had befallen him; and then
followed some years of hopeless misery, in which it
was not possible to find anything that could for one
moment occupy him or take his mind from what
he had lost. He had been a metal-carrier, and his
job had been day by day to lift the great logs of iron
when they had cooled and carry them to the stock-
yard: he had cared about nothing but the blast-
furnaces and their surroundings; he thought of
nothing else but his work; he had been absorbed
in it, and when he lost it he had nothing left.
Well-meaning people tried in vain to bring him
the solace of occupation, to make him do various
things with his hard hands, that had never done
anything since he was a boy but lift iron; they
tried to divert his mind, which, even in the times
when he could see to look at a printed page, had
never cared to see anything but his work. At last,
as the years went on, he became calmer and even
cheerful. He remained in the house he had always
occupied, a tiny four-roomed cottage in the very

middle of the works, within earshot of every sound from the furnace—not actually in want, for the man to whom such a misfortune happens is entitled to compensation at the hands of the law as well as of humanity. He had been a healthy and vigorous man, and if it had not been for this accident would probably have been able to go on working into old age. And all through he was tended with the most unceasing care by his frail, devoted old wife. She finally died before him, never having recovered from the shock to her nerves caused by his accident and the subsequent incessant strain upon her. After her death the daughter who remained to him took her place.

CHAPTER V

OLD AGE—JOINT HOUSEHOLDS—BENEFIT SOCIETIES

THE problem of how old age shall be encountered when the power of work is failing and resources diminishing, that dark prospect always in front of the worker, of the night coming in which no man can work, is to some degree dealt with in the community we are describing by giving a man who can do any work at all, or, indeed, anything that has a semblance of work, some lighter 'job,' and keeping him on, instead of turning him off after the years of long service during which he has grown old. There are, no doubt, some splendid specimens of health and strength to be seen among those who, inured to the life and to the variations of climate, go on working into old age. But the prospect in front of the man who has done this, and has worked all his life, unless in the rare cases of those who have contrived to put something aside for old age, is not a cheering one. The man who has had a strenuous and important post, requiring the utmost vigilant thought, is obliged, perforce, to accept some inferior job, that of a labourer, perhaps, at much less wages, and he is bound to be less competent at it than a labourer who is younger. Some men, it is true, in course of time become foremen, in which case their actual duties are less fatiguing, being for the most part supervision instead of actual work, and they are better paid ; but, of course, all cannot become foremen, and all are not qualified to do so : and there are many others who must perforce be content to accept, instead of the work hitherto accomplished, something for which so much strength, physical energy, alertness, and endurance is not needed. It is, no doubt, always a bitterness to a man to have to content himself with a lighter, less well-paid, and less responsible job than the one he occupied when he was hale and vigorous.

One very old man had worked for the same firm for fifty-seven years. For thirty years he managed to keep on with the furnace-man's arduous work, receiving therefor from £2 to £3 per week, and was then obliged to give it up. He was then given a job as weighman, for which he received 21s., sitting in the weigh-cabin all day and checking the wagons as they passed. Over and over again one finds such instances: a man is doing the work of a labourer, perhaps, in his old age, at 18s. a week, who has been in receipt of twice or even three times that amount when he was younger. Sometimes, of course, it happens that even this possibility is not left to the old, that the man, from broken health, debility, or disablement, is not able to do anything at all. In nearly every case this difficult question, which so many attempts are made periodically in this country to organize officially upon a settled basis, is empirically dealt with and practically solved by the old and enfeebled man being taken in and cared for by his younger relations. If it is an old man left quite alone, with no one belonging to him, he has to go into a lodging and manage as best he may, as a last resource applying for parish relief, which is kept at bay as long as possible; but it does not seem often to happen that the man in old age is as solitary as this. Few of the men have not been married, and those who have nearly all have ' belongings ' to look after them.

Over and over again one comes upon these joint households among the working homes. One finds a poor old man, whose declining strength is no longer equal to his work, living, if his wife is dead, with his married daughter or married son, and looked after by them. Or, what is perhaps more likely to be a sub-versive element in the household, the mother, if she has been left a widow, and cannot work or make a living by taking in lodgers, is taken in by one of the family. Sometimes even the husband and wife are both taken in if the husband is alive too, and cannot work. It is true that the result is not always all that can be desired; in many cases, as might be expected, there is friction, there are difficulties: the manners

of the inmates to one another are often unspeakable.
But even if their relations to one another may not be
ideal, there is, at any rate, no attempt made to shirk
these duties, and the younger people usually seem to
take it for granted that elder kinsfolk left alone and
unprovided for must be taken in and harboured.
Indeed, this claim of kinship is sometimes stretched
to a surprising extent. A visitor found on one
occasion in a cottage an old woman seated by the
hearth who had not been there before. On inquiry,
it appeared that she was a relation of the wife's, an
Irishwoman who, finding herself unprovided for and
alone, had thought of her young kinswoman, this
girl, whom she had not seen since she was a child,
and had started off from the other side of Ireland,
and suddenly appeared, telling the young couple she
had come to live with them. They appeared to have
taken it as a matter of course, and so far the arrangement seemed to be prospering. And yet, having an
extra unexpected permanent inmate in a house with
only four rooms means something very different from
what it would mean in a house containing twenty:
and even in the latter a joint household is not an
easy thing, given the customs and habits of this
country, to carry out with success. Still, there is no
doubt that in the cottages the households are often
rent by dissension, and that the older member who
has been harboured, feeble and ailing, it may be, is
querulous and complaining. This is more apt to be

the case with a woman than with a man, for the old woman is generally more outspokenly critical of the way the house is conducted. But many a time, on the other hand, the joint household is entirely and amazingly cheerful and contented.

Here is a case one cannot pretend is typical, but similar homes are not entirely uncommon. The father is a foreman at the ironworks; he has worked for the same people forty years. In the house live also the married daughter and husband and a daughter of three years old; the married daughter and a servant-girl do all the housework. The mother keeps a little shop. There also live in the house five sons, who are all employed at the works, and three more daughters, one teaching in a school, and two younger. This household seems most prosperous and united, and having so many bread-winners in the family, father, mother, and eight children, they are all very comfortably off.

Another delightful home was one in which a very old woman, a widow, who could not remember how long her husband had been at the works, but 'a great many years,' she said, lived with her daughter in a house of four rooms. The daughter had been a widow, and had two families. When they were visited, the younger woman was busily scrubbing her stairs and making them shine again, and trying to keep an eye at the same time on a very fretful baby, whom the eldest boy, about thirteen, was holding till the stairs were done. Her husband also had worked for the same firm for many years. The house was as clean and tidy as possible, the two women evidently on the best of terms. They said that there was always peace there, and that the husband of the younger one never 'touched anything,' for which they could not be sufficiently thankful. The wife seemed to take as a matter of course that she should 'do for' her mother and her children as well as for her husband and herself, besides having the rather ailing baby almost always in her arms.

In S. Street a workman, left a widower, lived with his mother-in-law, and had done so ever since his wife and baby died nine years ago, within a month of each other. He was lying ill in bed when a visitor called, apparently comfortable and well looked after, the room beautifully clean and tidy. The mother-in-law spoke of his great kindness to her, which she certainly seemed to return.

In V. Street another man, left a widower, came to live with his mother and share expenses of the house. This mother, also a widow, had taken in, besides her four grandchildren, the children of two married daughters who were dead. One of these grand-children, aged six, is a cripple.

In a house in R. Street an old man of seventy-three, who had worked thirty-two years for the same firm, was unable to do so any longer. He had two daughters and a son, one of the daughters married, the other, and the son, unmarried. The married daughter's husband fell into ill-health, and was unable to do anything; but they all agreed to live together, the married daughter looking after the father and doing the whole of the housework, the other sister going out as a laundress, the brother working at a foundry. For this household they had a house of four rooms in the very poorest quarter of the town, a small kitchen at the back, a small square parlour in front, and two rooms overhead. In the parlour, which was scrupulously clean, and adorned with china ornaments, etc., was a polished cupboard, which let down and formed a bedstead at night for the married couple. In the room of the same size above was a large bed in which the old father lay in his last illness, almost unconscious; in this bed, even at this stage, the son also slept at night. The unmarried daughter had the little room at the back over the kitchen.

In S. Place a young married couple, the man earning 28s. a week, took in the mother of the man, a widow aged sixty-six. They gave up to her entirely one of the rooms of their four-roomed house, rent free, and allowed her 1s. 3d. a week, and another son working somewhere else allowed her the same. But in this case the result was not so completely satisfactory; the old woman constantly complains, and tells every visitor who calls how badly she is treated.

On the whole, I think it more often happens than

not that the tie between the old parents and the children is one of strong affection as well as of duty. One constantly finds on going into a cottage where the young wife is ill, or her children are ill, her mother managing the situation, the competent motherly old woman sitting by the fireside with one or more of the babies in her arms, and seemingly steering the whole household. On one occasion the daughter of one of the workmen, who had been born in one of the little streets in the very centre of the ironworks, had married a man who was living in the other little settlement about half a mile off. She was one day found in floods of tears, saying that she missed her mother so dreadfully that she didn't think she could be happy so far away from her; it was so lonely, she said, when her husband was away at work, and she 'got studying' (*i.e.*, brooding) when she was alone.

The affectionate relation between the young married daughter and her own home, indeed, sometimes causes an additional difficulty, as there are cases where the young wife neglects her own house to go to her mother's.

A man in V. Street had had an accident, his legs had been scalded, and he had been seventeen weeks in the Infirmary, and at home afterwards for seven wretched weeks. When he began work again he felt miserably weak, as might be imagined, but his chief ground of complaint was that his wife neglected the house and children, and was always at her mother's instead of at home, leaving the husband to manage for himself. His two boys had been sent to live at the grandparents' during the father's illness, and remained on there.

At a house in W. Street the young wife shuts up her house all day and goes to help her married sister in her house, leaving her children to the care of her old mother, who looks after them very incompetently. The result is that the house is untidy, and the man uncomfortable and angry.

The story of another household was somewhat disastrous. The husband, a steady, respectable man, died, leaving two married daughters and a son, who unquestionably, albeit somewhat angrily, accepted the fact that they must provide for the mother, who stayed with them each in turn until they could bear it no longer, and then went on to one of the others, complaining of the one she had just come from.

But, on the whole, it is surprising to find how many houses there are in which, in spite of crowding, of provocation, of inconvenience, life seems to be conducted on sufficiently amicable terms.

And it is not only when there is a claim of relationship that the people we are describing are willing to assume the charge of an extra inmate; they often do so cheerfully and ungrudgingly when the new-comer is no relation at all, but simply a neighbour who is in trouble. When people who are better off have an extra person quartered on them, who is, be it said, more often a visitor than a neighbour in trouble, it merely means that the servants get a spare room ready, or that, at the worst, some of the family double up in a house where there is room to overflow. They can never in their lives hope to approach to the hospitality with which, without hesitation, the poor will give up some of their scanty room, and charge themselves with the entire burden of some one worse off than themselves. In one case the husband, an ironworker, had been ill with rheumatic fever and pneumonia, the wife with consumption—both hopelessly ill; the husband died first, and the kindly neighbour, a young unmarried woman who kept a little shop, offered to take in the dying woman, who shrank from going to the hospital. She took the invalid into her house, and, when the mother died, adopted the child.

Many of the houses take lodgers; it is the obvious way of arranging for the unmarried workmen. It is, indeed, sometimes a difficulty to prevent it where the accommodation is inadequate. It is disconcerting to find, when trying to arrange for better housing and more room, that in some of the houses there would be room enough if it were not for the lodgers. In a house of three rooms in M. Street, that of a labourer earning 25s. a week, there were actually found to be living the husband and wife, two

boys, two girls, and five young men lodgers, all
employed at the works. This, of course, is an
extreme case, but it is not by any means unique.

In another house containing three rooms the
parents inhabited one, three children another, and
the third was let to two lodgers who were employed
on different shifts, and who used often to go into
the room and to bed without the room having been
touched or the bed remade. But this, also, although,
unhappily, not a solitary instance, must not be taken
as typical. When old workmen have no belongings
of their own to look after them, they are taken in as
lodgers at one of the houses at the works, and often
looked after with care and solicitude; but at the
best, except in the rare cases where it has been
possible to make some provision for it, old age,
which, if only for the mere lapse of the years, presents a melancholy aspect to so many of us, must
indeed look dark when it represents either destitution
or dependence.

We are apt to be surprised, when we consider the
constant risk of illness and accident to which the
ironworkers are subject, as well as the prospect
before them in old age, to find that by no means all
of them either insure their lives or put into any form
of sick club. And the fact that there are so many of
even the sensible and respectable workmen who do
not take either of these precautions is one of the
salient patent facts that we are most ready to label

as want of thrift. Out of the first 700 who were
interrogated on the subject, 380 were in a club, 80 in
two clubs, 270 in none. Further investigations,
bringing the number up to over 1,000, have confirmed these proportions. On the face of it, considering that the disastrous consequences of being
off work through sickness is a fact which must be so
constantly before the workmen, either in their own
experience or that of others, these figures at first
seem almost incredible, when we remember that
those who have had the forethought to put 6d. or 9d.
a week into some one of the many sick clubs receive
10s. or more a week during illness, making it possible
to tide over the time until they go to work again.
But on further consideration it becomes not only
credible, but probable : and the reason of it is not
necessarily that the man has deliberately and per-
versely declined to make provision for such a misfortune, it is very often that, like the rest of us, he
is 'just going' to do it ; he keeps thinking while he
is well that he will do it, he meant to do it yesterday, he means to do it to-day, he will certainly do it
to-morrow—and then suddenly, when the moment
comes, he finds too late that he has not done it at
all. In fact, in many cases the workman does not
put into a sick club for the same reason that his
employer does not take out his umbrella when the
sun shines—because, both literally and metaphorically, when it is fine it is difficult to realize that

it will ever be wet. And belonging to a sick club
means taking definite steps beforehand; there must
be a medical examination, there are forms to fill up.
Many of the men are too shy to go up for the medical
examination, some do not know where to apply, and
do not know whom to ask for information. Many
among them have a quite insurmountable aversion
towards embarking upon anything which would
necessitate coming in contact with officials, filling
up forms, etc. It is essential to realize, when one is
engaged in any undertaking connected with the
working classes, what a very great gulf there is, and
must be, between their methods of approaching it
and those of the well-to-do, who are accustomed in
any such contingency to bring into play the whole of
the machinery available to them in the developments
of modern life, to use the telegraph and the telephone. It does not occur to them, even if they could
afford it, to use any of the means of communication,
any of the abbreviated forms of intercourse, which
are as a matter of course to their employers. On
one occasion it was necessary that a visitor should
meet a workman at his own request, transmitted to
her by a verbal message through another person, to
discuss a question which had to be settled at once.
The visitor did not hear from him, and did not know
what had happened until she met him casually in
the street; he then told her of an urgent message he
had been anxious to deliver to her for about a fort-

night, but had waited until he had happened to see her, about getting his wife into a hospital at once.

The same vagueness as to knowing 'how to set about it' applies to other organizations for promoting thrift. There are the Penny Savings Bank, the Post Office Savings Bank, and others, all a good deal used; but all of these imply some definite step to be taken. The process of saving is made easier, as far as the formalities surrounding it are concerned, by some employers, who offer to take charge of the men's savings and give them 5 per cent. on them, with no need for any steps to be taken by the men, since the sum is simply stopped off their wages if the men ask to have this done. A certain proportion of them are glad to do this; others, even among those who are thrifty and willing to save, do not wish to do it in that way. There seems to be among some of them some obscure feeling that the employers have an extra and unlawful hold over a man when they know what balance he has at his bank; it is going into his very stronghold. It would perhaps be putting it too definitely to say that this attitude on the part of the workman to his employer—and not only on the question of savings—amounts to suspicion or mistrust; but it is distinctly guarded. He waits to see. It is the outcome of a shrewd, sensible wariness, which asks itself, when a given step is taken by somebody else, what the likely reason is for taking it. That shrewdness on such an occasion is perhaps too little apt to assume that sentiment or a care for the welfare of others might be a possible motive of conduct, although not necessarily the only or the invariable one.

As has been seen from the figures given above, more than half of the workmen do put into sick clubs; some, indeed, belong to more than one. There was a man in X. Street who, out of a wage of 36s. a week, paid 9d. a week into the Hearts of Oak Club, and if off through illness was entitled to draw 18s. a week. He also paid into a bricklayers' club 8½d. per week, and if out on strike could draw 15s. a week. Another man in S. Road was in the Joiners' Association Club; he paid 1s. 9d. a fortnight, and drew 10s. a week if off work ill, 15s. if on strike. This man also insured his tools for £40 against fire and misfortune. More than three-quarters of the men employed in the works described belong to trade unions; these men get 10s. a week from their union in case of sickness.

There are, in the town in question, nearly 200 branches of various friendly societies. Many of these meet at public-houses, and thus give the opportunities for conviviality that may be imagined. Some of the yearly benefit clubs of which the head-quarters are at public-houses demand, besides the weekly subscription agreed upon, an extra contribution, from 1d. to 3d., for the good of the house, what is called the 'wet rent,' which is quite deliberately allowed for drink each meeting-night. The amount thus obtained is divided among the members present at the yearly meeting; thus, supposing there were a membership of thirty, and that only six men attended the meeting, that would leave thirty times 3d.—that is, 7s. 6d.—to be divided among them in drink. These are not ideal conditions for the encouragement of thrift, and many a time a man who is trying to do his best for his family by putting away a certain sum of money a week is, strangely enough, doing it under conditions which, while they give him the opportunity of thrift, at the same time encourage drinking and extravagance. The landlords, naturally, are willing enough to have their houses as headquarters of the club, as it attracts custom. Sometimes a bottle of wine or spirits is given to the members at the yearly meeting.

Some of the more sensible among the workmen, seeing the drawbacks to the benefit clubs that meet in public-houses, have started clubs on their own account. The way that such a benefit club is formed is as follows: The names of the people willing to be members having been obtained, and the subscription decided upon, each man pays an extra fortnight's subscription as his entrance fee, besides the sub-scription itself. For the first six weeks the club pays out nothing. After that every man, when he is absolutely off work through sickness, receives, say, 10s. per week. Upon his death the wife of a member receives £5, he himself, if the wife dies, £2 10s. If anything happens to be left in hand at the yearly settlement, it is divided among the members who belong at the time. It is an almost invariable rule, in order to minimize the danger of malingering, and to add to the difficulty of pretending indisposition, that there are no payments for the first fortnight after membership. If a man is ill at the breaking up, his pay ceases until the next year's club, a rule which often falls heavily upon the members. They elect from among themselves the officials of their society, which consist of a treasurer, a secretary, and a sick visitor. The function of the latter is important, as he goes round to ascertain that the people who apply for relief to the sick fund absolutely are on the sick list, a process which sometimes leads to heart-burnings among the people visited. I have heard a man complain that 'if you so much as take your baby on your knee you are supposed not to be ill enough for benefit.' Another was indignant because he was found lifting a kettle from the fire, and was therefore pronounced not ill enough to be entitled to the benefit of the money.

Many people who do not succeed in making up their minds to put into a club, start, with good intentions, saving at home. It is less trouble to put the available surplus into a box or bag in the cupboard indoors than to go out of the house and take it somewhere else. But the money kept in the box or bag in the cupboard is pretty certain not to remain in it long. And there comes in the practical difficulty of accumulation, the want of a suitable and safe place to put the money in, the additional difficulty of keeping it, if it is on the premises, by the knowledge that it is there, and can be taken from in case of need—taken for the number of opportunities of spending money which present themselves to the working-man, as to every one else, at every moment of the day.

CHAPTER VI

RECREATION

WE must first of all reckon with the inevitable, permissible, and almost universal craving for diversion, for the change of thought and scene, which, unhappily, for nearly every one must be bought for money. What are the opportunities for such change open to the people we are considering ? In the summer, perhaps, these are plentiful enough, especially for those who can bicycle and who can go far enough from their usual surroundings to get in a very short time into something like country. Others throng the trams and the trains. One of the station-masters of the district told a visitor recently that it was quite amazing to see the number of workmen who were constantly travelling to and fro. It means, after all, they share the general restlessness of the community. But this restlessness and desire to travel, to move, has increased to an extraordinary degree. One man who had 36s. a week spent 17s. of them on his expenses to a town some distance off, in cold, unfavourable weather, to see a football match on which, according to his own showing, he did not even care to bet. But it was somewhere to go and something to think of outside his own daily work and surroundings. There are not many opportunities in the colder months for what may be called out-of-door amusement except watching football matches, a comfortless thing enough to do in a North of England winter. It must be said, however, that the cold does not seem, on these occasions, to damp the enthusiasm of the onlookers. Thousands of spectators will watch these matches, the excitement of many of them, no doubt, whetted by having a bet on the result, but many of them watching out of sheer interest in the game.

Where else has the workman to go? He obviously, during most of the year in the climate of the North, wants a place under cover; he wants a place which is warm and light and bright and cheerful. Where is he to find it? There are, it is true, several work-men's clubs in the town, opened by certain firms as private undertakings for persons in their employment; these are eagerly used by those entitled to do so, but there are not many of them. I am perhaps well within the mark in saying there are only about a dozen of these clubs, as opposed to 168 premises licensed for the sale of intoxicating liquor. There are, as possible places of resort in the town, two theatres and two music-halls, both well patronized, the latter especially always crowded; but these, except two days in the week, are only open during the evening, and, even if always desirable, are an expensive form of amusement. There is a very admirable Free Library, in connection with which is a news-room where the daily and weekly papers are to be found. This room is nearly always full of readers, many of them working-men. There is also an interesting museum, but even though this is much frequented, there are still many workmen whose spare time is unprovided for. Those who do not know enough to enjoy the Free Library or the museum—and these are many—who do not wish to spend their money on the theatres and music-halls, and who are not entitled to go to any private clubs, have nowhere to go for change of thought and diversion but the streets, unless they turn into the ever-present, ever-accessible public-house, where they find society, conviviality, amusement, where they can enjoy looking on at various games or taking part in one themselves. Billiards is the most popular of all indoor games, and in all the public-houses and in nearly all the private clubs tables are to be found. Billiard matches are constantly going on, interesting to watch; or it is possible for a man with a penny in his pocket to enjoy a game himself. If he loses he pays twopence—that is, he pays both for himself and his adversary. To judge by the way the tables are surrounded, and by the number of people waiting their turn, there are many men who have this twopence available. In cases where the man, by arrange-

ment with his wife, has his own pocket-money, this
comes out of it. He has therefore to pay twopence
for twenty or thirty minutes' diversion. There is,
I believe, a widely spread idea that it is better that
the workman should always have to pay for his
billiards, even in workmen's clubs of a more or less
private nature. I feel doubtful about this, since if
the loser pays for both, as is generally though not
invariably the case, it is practically playing for money,
though sometimes, however, the expense is divided
between the players. The possible enjoyment of a
game of billiards therefore becomes more or less the
monopoly of the skilful players, who, since presumably
they become by practice more expert, become more
difficult to oust. On the other hand, if the billiards
are free in a club to which the workmen contribute
in another way, that is, by paying a fixed sum weekly,
it means that a much larger number will have the
opportunity of playing, that there will be a constant
possibility of recreation to men who cannot and ought
not to afford to pay twopence for one game of billiards.
Instead of being a pleasure snatched with misgiving
and partaking of gambling, open to comparatively
few, it will be a safe recreation which is eagerly taken
whenever the tables are free, and open to many who
would be otherwise debarred from it. This may not
be so exciting to the expert billiard-player, whose
interest is whetted by the sense of playing even for
a small sum; but we are not considering him at this

moment. We are considering the large number of
others who, unless the tables are free, have no opportunity for playing billiards at all. The advantage of
the free table is not an imaginary picture. I have
seen a club with two free tables, where men who,
being either on the afternoon or night shift, have
their mornings on their hands, have been happily
knocking the balls about from 9 a.m. onwards, at
a time when there are few people in the rooms and
the more practised billiard-players are not frequenting
them.

Many a workman, whether in his own house or in
a lodging, finds the hours when the housewife is busy
and the house upside down a comfortless time enough
to live through if he has nowhere else to go. And
not only in the morning, but all day and every day,
this is the permanent condition of many a house
where the woman is not skilled in the domestic arts,
and comfort is unknown. It must also be borne in
mind that such a man as I am describing might presumably, if he were not knocking the balls about, be
at the best loafing at a street corner.

The objection sometimes made that it is difficult
if the billiards are open to all to regulate the succession of the game is really not insuperable, as it can
be obviated by names being put down for games
beforehand and the games being strictly played in
order. In all clubs there is presumably some one in
authority acting in conjunction with a committee,

who could see that this regulation was carried out.

As far as the women are concerned, there is still less provision made for their diversion and recreation than for that of the men, except by private undertakings of one kind or another; as to public places, there are not many in this country where it is the custom for ordinary domestic respectable working women to go, and it is much less common, therefore, to see a workman taking his wife and family out pleasuring with him than it is in most countries on the Continent. It all means, I imagine, that there is less definite provision for the occupation of the workman's leisure in this country than in most others. And yet this is the thing for which there is an urgent need.

In these days there is, in all classes, an increasing and often unreasonable demand for pleasure, a sense of irresponsibility towards the employment of time. But so far as the workman is concerned, it is a good deal to demand of the average man whose work is of a strenuous physical kind that he should be equal to a mental strain in his hours of leisure, or that he should have energy to invent or create occupations of a desirable kind for himself; if we do demand it, we are likely to be disappointed. The resources provided for a man's leisure matter incomparably. It is during these that he may be ruined and dragged down, and not in the hours of his work.

The need of some provision for these men's leisure is especially conspicuous on a Sunday. Sunday is the workman's day of leisure, and he wants a place to go to on that day where he will be amused. An investigation recently made by a local temperance society (the ' church and public-house census ' already quoted on p. 10) into the numbers who entered the public-houses in the town on a given Sunday, gave the following result. Into 106 public-houses and 36 ' off-licenses ' observed that day, there entered :

Men	...	...	...	55,045
Women	...	...	...	21,594
Children	...	...	...	13,775
A total of ...		...	...	90,414 persons.

The majority of these people probably belonged to the ironworkers.

As stated in an earlier chapter, there are 70,000 people who do not go to church, a large majority of these belonging to the working classes, and of these, again, probably a considerable number would like to find something attractive to do on this day of leisure outside their homes. The workmen's clubs are not open on Sunday, the theatres and music-halls are not open on that day, nor is the Free Library, nor the museum. The public-houses are open, and are therefore frequented by the numbers shown.

If it were possible that there should be scattered about the town various places of resort under cover, open during the winter at an almost nominal charge, places well warmed and lighted, open to anyone and every one who chose to pay, where a man might turn in and sit down, have his pipe, and meet, during his free hours, with his fellows, it would, I believe, make an incalculable difference to the welfare of the community. They might be places of a kind to which the women might go too. It would make a difference to many an overworked wife and mother if she could go out with her husband in the afternoon and have a cup of tea, perhaps, which she had not made herself, in some warm and pleasant surrounding. There might be refreshments provided at cost prices, and when it was possible some music from a band or even a piano.

It must be remembered that the workman who wishes for a change of thought and scene after a day's work is practically debarred from much of the social intercourse of a private kind which plays so large a part in the existence of the better off, who are for ever going to dinners and parties at one another's houses. He obviously, in most cases, has no place in which to receive his friends, if he wished to do so. In nearly all the cottages, if there are many children, the available space is all taken up. The women do go to see one another: you will often find them sitting in each others' houses. More rarely a man who has called to see one of his friends, may be found sitting in the latter's house ; but there can, of course, be no organized, enjoyable, entirely recreative social meetings in individual houses as there are for the better off. The workers depend entirely upon the opportunities of reunion and of recreation either that are organized for them or that they organize for themselves in common. In the settlement of cottages on the north side of the river, in the midst of the works, weekly 'socials'—*i.e.*, dancing-parties—are got up by the workpeople themselves, directed by a committee of the more respectable inhabitants. These are held in the evening in one of the schools, and the payment is threepence each, which also covers the cost of refreshments.

The other places of entertainment open to the workman are the theatres and the music-halls. In the two music-halls in the town, which are always full, the dearest places—excepting the boxes, to which, apparently, only a select public go—are 1s., the price of the orchestra stalls. The dress circle is 6d., the pit 4d., the gallery 2d. In the gallery there are always a great number of boys, as well as in the pit. The front row of the gallery generally consists of small children, little boys between seven and ten, eagerly following every detail of the entertainment. Each of them there must have paid 2d. for his place—how he acquired it who can tell? probably either by begging or by playing pitch and toss in the street. There are workmen to be seen in the orchestra stalls; that means 1s. a night. If a man

takes his wife with him that means 2s. : but there
are many more men than women to be seen there.
Women go oftener to the cheaper places : one may
see a ' queue ' of them waiting to go to the 2d. seats,
often with their husbands accompanying them.
Many of these women have their babies in their
arms. There is no doubt that they come out
looking pleased and brightened up. The kind
of entertainment usually offered does not, to the
more critical onlooker, seem either particularly
harmful nor specially ennobling. The curious fact
that, in almost any social circle, it makes people
laugh convulsively to see anyone tumble down, is
kept well in view and utilized to frequent effect.

As to the theatre, the stage in such a community
as we are describing—or, indeed, in any other—has
an immense opportunity. The stage is at once the
eternal story-teller and the eternal picture-book. The
repertory of the two theatres in the town, fortunately,
does not consist only of reproductions of London
successes of the most trivial kind. These are occasionally performed, but more often the plays are
sensational pieces of a melodramatic kind—that is,
usually sound and often interesting plays, in which
the boundary of what is commonly called vice and
virtue is clearly marked—virtue leading to success
and happiness, vice to a fate which is a terrible
warning. For my part, I wish that such repre-
sentations, such pieces as these, could be multiplied,

that they could be constantly accessible at entirely
cheap prices for the ironworkers and their families—
indeed, for the whole of the population. I would
like to see some building of the simplest kind in
every parish in which they could be performed.
There is a small town a few miles distant from
Middlesbrough to which there comes at intervals a
stock theatrical company, which performs literally
in a barn, at infinitesimal prices. The plays produced, if not very nourishing to the more complex
mind, are always sound and good, full of movement,
full of interest to the audience before whom they are
performed. Night after night that barn is full; night
after night men and women, boys and girls, who
might be loitering in the streets or in public-houses,
are imbibing plain and obvious maxims of desirable conduct, are associating mean, cowardly, and
criminal acts with pitiable results. No one who had
been to that little theatre could doubt the good effect
of the influence that must be radiating from it, and
it would be well if such centres of influence could be
found in every manufacturing town. There are, no
doubt, many people, and some of these are to be
found among the working class, who disapprove of
this medium of entertainment. But there are also
many of us passionately convinced that the stage, if
used in this way, would be an influence more for good
than for evil; that it would offer countless opportunities of suggesting a wholesome, simple, rough-

and-ready code to many listeners who since they left
school have probably not had any moral training at
all, and the majority of whom are more than likely
to drift along through their lives at the mercy of
every passing influence.

The ordinary man is presumably the creature of
his training, his surroundings, his inherited tendencies. There are plenty of men among the ironworkers, as in any other section of society, with
characters strong enough to fashion their destinies
for themselves, instead of being fashioned by them ;
but these are not in the majority. So far as the ironworkers are concerned, it is somewhat unreasonable
to expect to find special force of character in a layer
of society where, as has been shown, so much less is
done during the years that character is being formed
to bring it out. It would be more reasonable, even
if against the teaching of experience, to expect from
educated boys, who have presumably had, at any
rate, the opportunity of being taught worthy principles of conduct by precept and example, that they
should have the virtues we demand from the workman, rather than that the average workman, who
during the impressionable years that matter most has
had no deliberate training in ethics or morality at all,
should develop strength of mind and character under
pressure of need. For the workman's bringing up,
his influences and surroundings, have not been those
which tend to promote in him resolution, consistent

purpose, prompt action; on the contrary, he is more likely to be lacking in these qualities than are some of those to whom they are less essential. The boy of the ironworking district, when he leaves school, at the age of fourteen at latest, is in a part of the world where the principal industry offers hardly any occupation for boys. He is therefore, between the ages of thirteen and sixteen—that is, at the age of all others when, if he is to be a worthy man, a boy ought to be under supervision, under direction, when he ought to be given occupation and exposed to good influences—simply turned loose, either to do nothing, or else to take on one odd job after another of a temporary kind leading to nothing, running errands, selling newspapers, until he is old enough to take a job at the works, for which he is usually unskilled. He is mostly, as far as moral training is concerned, left to himself. The children of prosperous parents at this stage have their lives organized for them by other people. During the holidays their parents, and, indeed, many other members of the household, bend their energies to finding occupation for the spare hours of the schoolboys. When at school they are so carefully looked after, both in their school hours and their playtime, that they often complain of having to join the school games when they are not inclined to do so. Less fortunate boys have nothing to do with their leisure but to go about the streets, at the mercy of any temptation that may come in

[See pages 21 and 22.

GANTRY, BUNKERS AND KILNS.

[See page 32.

MINE-FILLER DRAWING IRONSTONE FROM KILN.

[*See page* 34.

CHARGING THE FURNACE.

[*See p*

MOLTEN IRON FLOWING INTO "LADLES."

their way. Many of them would be only too glad,
as various admirable experiments in vacation schools
and other directions have recently shown, to take up
any occupation which is offered to them. There are
various associations, boys' clubs, etc., which help
them to find occupation at given hours, especially in
the evening. Many of the workmen, be it said, take
a keen interest in these boys' clubs, and are invalu-
able in superintending them. But the boys still
have a great deal of time in which they loaf about
the streets with nothing to do. No doubt a certain
proportion of these boys are happily made of the
stuff which, in spite of all adverse circumstances, will
turn into worthy men—boys who are steady, anxious
to do well, and not to get into bad habits; but, as
a rule, I fear it is incontestable that most of the
children who are playing about the streets of
Middlesbrough are destined to grow up into a
generation which will bring down the average of the
deserving and efficient. This immense population
of workers is growing up among physical and
moral influences which are bound to be unfavourable.

The parents among the working class, even if they
have no special occupation to offer the boy, are
generally not at all anxious that he should remain at
school; in fact, many of them take him away the
moment he is thirteen, the age at which they are no
longer obliged to keep him there. A clever boy,

indeed, may have passed all his standards by the time he is twelve. One constantly comes across parents who, if their boys remain at school until fourteen, deprecate the fact, on the ground, they say, that the boy is likely to get into loafing and unsettled habits when he has passed through all the standards and got to the top of the school, and the lessons make no further demand upon him. Unhappily, even if this be true, the alternative does not seem to be much better. I remember one such instance, of which only too many could be given. The father, who was a workman in receipt of 35s. a week, took the child away from school on his thirteenth birthday, in the middle of the school term, for no reason at all. The boy for a year afterwards had hardly anything to do. He sometimes ran errands, but more often he loafed about the streets, with intermittent bouts of work, until he was sixteen or more. And yet we are going to demand from this boy and his fellows, arrived at manhood, many qualities—stern self-denial, consistent purpose, unceasing industry, self-sacrifice, temperance in its widest sense—not always, in spite of our well-meaning attempts both at home and at school to induce them, found either in boyhood or in manhood in the sons of the well-to-do. And it must be remembered that the latter have, during the years that they are at school and college, that is, during the ten years which begin at the age at which the schooling of the workman leaves off, had the opportunity of becoming acquainted with the works of the great writers, of the thinkers of every age, the philosophers, the moralists; and even if in later years the writings are forgotten, the study of them has often, presumably, leavened thought and character, if it has done no more.

CHAPTER VII

READING

In these days when books are cheaper and libraries more plentiful, the workman, no doubt, has more opportunities for study than he used to have : a great many more books are accessible to him than in former times. There are a certain number of born readers among the workpeople in the town described, as there are, happily, in every layer of society, who not only devour omnivorously the books that come under their hand, but also go further afield and find more for themselves. But these are but a small proportion of the total number. Reading, perhaps, is not so prevalent a habit in any class of society as we try to think; but there is such a pressure of opinion in favour of it that it is difficult for the most candid to admit that they do not care about it. All the same, I believe that these are in a much larger proportion than we imagine. Learning to read does not necessarily lead to the enjoyment of literature. It is, no doubt, an absolutely necessary step in that direction, but I cannot help thinking, on looking

at the results all round, and not only among the
workmen, that the knowledge and practice of reading
make nearly as often for waste of time as for edifica-
tion. We are apt, even those among us whose
habitual reading by no means constitutes a liberal
education, to exclaim with genuine surprise and
dismay when we realize what the working classes are
in the habit of reading, or when we find that they
do not read at all. We then solace ourselves by
bringing out one of those comforting easy phrases
which we carry about ready to apply to any place in
the social fabric that seems to need it, and we say :
' With the spread of education the working classes
will read something better worth reading.' But
will they ? The spread of education has a broad
back. It is made to bear the burden of many
unrealized, if not unrealizable, projects. Education is
being spread very thin indeed for the people who
are in question, and I will venture to say, judging
from what I have observed of the reading of men
and women who have been what is called ' educated,'
and of those who have not, that it is not the spread
of education that will alter the reading of the great
mass of the community, those who swell the tables
of statistics and bring down the average of the
enlightened. On the whole, I think it may be
stated that a large majority of people of both sexes,
in every walk in life, read, with hardly any selection
of their own, what comes under their hand, what is

suggested to them, and what they see being read by the person next door.

The reading that comes under the hand of the workman consists chiefly of the newspapers hawked about the streets, and those supplied by the small composite shops found in the poorer quarters. These shops, which sell various other goods—groceries, haberdashery—put before their public an unfailing supply of daily and weekly newspapers suited to their tastes, and penny novelettes. Roughly speaking, more than a quarter of the workmen read books as well as newspapers ; nearly half of them read the paper only, and a quarter do not read at all. Of these, some only read a daily paper, the favourite being a local halfpenny evening paper, which seems to be in the hands of every man and woman, and almost every child. It contains a summary of general news, a serial story, a good deal of sporting information, also gossip and commercial news. Many workmen read in addition to this paper the weekly Sunday papers, containing several sheets, and providing a good deal of miscellaneous information. These are great favourites, and help to make the Sunday pass quite harmlessly, at any rate, for many among the workmen, who spend the day in bed, reading and smoking. The Sunday papers are a very special feature in the literature of the working classes. They are provided in view of the fact that the Sunday is the principal day on which both men

and women are likely to read, and they consist of
special papers, such as *The Umpire, The People,
Reynolds's Weekly, Lloyd's News, Weekly Despatch,
The Week, News of the World, The Sunday Chronicle.*
And it is interesting to note that, even in households
where each penny is an important item of expenditure, 1d., 2d., 3d., and sometimes as much as 6d., is
set apart for this delectable Sunday reading.

As might perhaps be expected, the workman reads,
as a rule, more than his wife, not only because his
interest on the whole is more likely to be stimulated
by intercourse with his fellows widening his horizon,
but because he has more definite times of leisure in
which he feels he is amply justified in ' sitting down
with a book.' About a quarter of the men do not
read at all: that is to say, if there is anything
coming off in the way of sport that they are interested in, they buy a paper to see the result. That
hardly comes under the head of reading. The boys
read papers that make them laugh, *Comic Cuts* and
the like. I have seen a large number of comic
illustrated papers, compared with which *Answers* and
Tit-Bits are the very aristocracy of the press. The
Winning Post and other sporting papers, which
represent the whole world as governed by the huge
spirit of betting, are more and more in the hands of
readers of all ages, with very undesirable results.
The question of betting literature, for which there is
an ever-increasing demand, is too wide to go into

here. The *Illustrated Police Budget* is a sensational and much-read paper. In a number which has lately come into my hand there is a special double-page illustration headed 'Father Murders Six Children.' Outside the page, on one cover, is a picture of a man cutting his wife's throat, on the other of an actress being thrashed by an irate wife—the counterpart, in a cruder form, of the detective stories revelled in by readers of more education and a wider field of choice, such stories as 'Monsieur Lecoq' and 'Sherlock Holmes.'

The above publications, which make so large a part of the reading of the men and boys, are something between a pamphlet and a newspaper. Their reading of books comprises novels, sometimes travels, hardly any poetry, a few essays, books relating to their work, and one or two biographies.

I give here, taken haphazard and, where it is possible, in the very wording, the replies made in 200 houses visited to questions respecting the reading of the inmates. These replies speak for themselves. The weekly wages are inserted, where known, as a guide to the standing of the man; the calling of the men quoted has not been given for fear of identification. They are all people employed in various capacities at the ironworks.

1. Husband and wife cannot read. Youngest girl reads the paper to them sometimes, but 'she has a

tiresome temper and will not always go on.' (Wages, 18s. per week.)

2. Does not care very much for reading—'just the newspapers.' (Wages, £2 10s. to £3 per week.)

3. Fond of reading.

4. Tries to teach himself German.

5. Husband reads newspapers only, particularly racing news. Wife spends much time in reading penny dreadfuls, illustrated papers which she considers 'thrilling, as they give such a good account of "high life and elopements."' Husband disapproves of his wife's tastes.

6. Wife very fond of reading, particularly children's books. Husband dislikes reading anything.

7. Likes reading the Parliamentary news, and doesn't approve of 'local nonsense.' Wife fond of reading stories of country life. (Wages, 25s. per week.)

8. Fond of reading, especially history. (Wages, 30s. per week.)

9. Great reader, particularly of history, 'sea battles and their results.' His wife reads Shakespeare and revels in 'Hamlet.' 'For lighter reading' likes Marie Corelli and Miss Braddon. (Wages, 36s. per week.)

10. Very fond of reading books relating to his work, which he borrows from his brother. Wife does not care for reading. (Wages, 37s. 6d. per week.)

11. Great reader, saved up his money to buy engineering and theological books. (Wages, 28s.)

12. Brother does not care to read. Sister very fond of reading, and reads Mrs. Henry Wood and Shakespeare to the rest of the family.

13. This man's wife never learnt to read, as she was brought up in some remote country village, and the 'school-dame had a spite against her.' (Wages, 9d. per hour.)

14. Reads everything he can get hold of. (Wages, £3 per week.)

15. Very fond of reading books relating to his work. (Wages, 36s. per week.)

16. Husband cannot read. Wife only time to read on Sunday, and then reads the *Sunday Magazine*.

17. Great reader, particularly seafaring tales. Wife does not care to read except on Sundays, when she prefers the *News of the World*, which she considers 'a great paper for enlarging the mind.'

18. Great reader; prefers above all to read of wars, in which he takes the greatest interest. He has 'often been mistaken for a soldier from his conversation and great knowledge of battles.' Wife very energetic woman, doing her own baking, washing, papering, etc., and no time to read. (Wages, £2 10s. per week.)

19. Fond of reading romances. Wife does not care for reading at all, prefers sewing.

20. Only cares to read about sports, anything else he considers trash. Wife cannot read.

21. Husband cannot read. He says when he was young he was tongue-tied and no trouble was taken

with him; he still has impediment in his speech. Wife always reads to him in the evenings, and by this means keeps him at home, but he will listen to nothing but romances.

22. Husband's chief delight is in reading detective stories. Wife cannot read.

23. Neither husband nor wife can read. The wife was brought up by an aunt who would not allow her to go to school. Her boy is going to teach her to read.

24. Fond of reading 'downright exciting stories after his work is done to get his mind into another groove.'

25. Husband *very* fond of reading—novels only—so is his wife. (Wages, £2 10s. to £3.)

26. Both husband and wife are great readers—novels—which they get from the Free Library and borrow from their neighbours. Husband reads aloud to wife. (Wages, 3s. 3d. to 3s. 9d. per day.)

27. Neither he nor his wife can read. (Wages, 21s. to 25s. per week.)

28. Reads newspapers, evening paper chiefly. (Wages, 25s. per week—varying.)

29. Does not read much, as his eyes were burnt in an accident at the works. Wife no time to read, and prefers to 'be lazy on Sundays.' (Wages, £3 per week.)

30. Great reader, and has quite a collection of books on various subjects, which he has bought cheaply at the market.

31. Neither he nor his wife care in the least for

reading : they occasionally 'pick up a paper.' (Wages, 30s. per week and overtime.)

32. Reads newspapers only.

33. Is devoted to books, spends all his spare time reading. Is a member of the International Library of Famous Literature, and has a small bookcase full of splendid books which he pores over every night to his intense enjoyment. He also gets books from the Free Library on science, blast-furnaces, etc. His wife cannot read and refuses to learn, although he is most anxious to teach her. (Wages, 5s. 11d. to 6s. 3d. per day.)

34. Husband cannot read, but likes pictures, and his wife reads to him.

35. Neither he nor his wife cares for reading : occasionally they 'pick up a paper.' (Wages, 23s. per week—varying.)

36. Is fond of reading and interested in Parliamentary news.

37. Is a great reader, and has 'taught himself mostly, as he was only at school two years.'

38. Reads to his wife, as she is nearly blind. (Wages, 4s. 9d. per day.)

39. Both fond of reading—husband, sentimental stories; wife, biographies. (Wages, 4s. 6d. to 5s. 6d. per day.)

40. Not readers. (Wages, 5s. per day.)

41. Husband reads sporting papers only; wife, boys' tales of adventure.

42. Neither fond of reading.

43. Reads a great deal at the workmen's club

to which he goes, but will not bring any books home.

44. Reads aloud to his wife, chiefly Mrs. Henry Wood's books.

45. Both fond of reading nice novels. (Wages, £2 10s. per week.)

46. Reads paper only. Wife can neither write nor read. (Wages, 26s. per week.)

47. Is a very great reader. (Wages, 26s. per week.

48. Talks intelligently about current events, and reads all the newspapers he can get hold of to his wife at home.

49. A great reader of fiction; wife would be, but has not time. (Wages, 3s. 4d. per day.)

50. Wife fond of reading exciting novels, which the husband considers waste of time. (Wages, 5s. 6d. per day.)

51. Wife fond of reading, but comic papers only. (Wages, 23s. per week.)

52. Wife never reads anything—all her spare time is taken up sewing. (Wages, 21s. to 23s. per week.)

53. Husband reads daily paper to his wife, and enjoys a good novel.

54. Great reader of seafaring tales. Wife no time for reading.

55. Never reads. Sons fond of reading papers— sporting news. (Wages, 5s. 3d. per day.)

56. Does not care for reading. (Wages, £2 10s. to £3 per week.)

57. Both fond of reading novels. (Wages, 4s. 9d. per day.)

58. Whole family fond of reading, especially fairy tales. They all read aloud in the evening, and the children repeat the stories to their playmates. (Wages, 4s. 9d. per day.)

59. All fond of reading. Wife likes Mrs. Henry Wood; the men prefer travels. Nephew is a good French scholar.

60. Both fond of reading—wife, novels; husband, chemistry.

61. Do not read.

62. Never read books. Husband reads daily paper. (Wages, 30s. to 35s. per week.)

63. All fond of reading. Boy reads books of adventure.

64. Read paper only. (Wages, 25s. per week.)

65. Both fond of reading, especially the *Sunday Magazine*. (Wages, 23s. per week.)

66. Both fond of reading, and prefer novels. (Wages, 36s. per week.)

67. Mother and son both fond of reading—she tales of adventure, he novels.

68. Daughters fond of reading: prefer novels.

69. All fond of reading books of travel and adventure. Husband prefers Parliamentary news.

70. Family do not care for reading.

71. Both fond of reading.

72. Fond of reading novels.

73. Not great readers; wife prefers knitting and sewing. (Wages, 50s. per week.)

74. Do not care for reading.

75. Husband fond of fiction. Wife likes sewing best.

76. Mother cannot read. Son prefers newspapers.

77. All the men, three lodgers, are fond of reading, and get books from the news-room library.

78. Do not read much ; think it a waste of time.

79. Goes frequently to Free Library and reads all the better magazines, also Plato, Aristotle, etc. ; considers their works ' such' splendid food for man's mind.' Has learnt a little French.

80. Prefer daily papers.

81. Fond of reading novels.

82. Both fond of reading.

83. Read very little, chiefly newspapers.

84. Not a reading family at all.

85. Reads books and magazines relating to his work. (Wages, £2 15s. weekly.)

86. Buys books referring to his work. (Wages, £3 10s. per week.)

87. All the family are fond of reading. (Wages, £3 15s. per week.)

88. Great reader. (Wages, £2 per week.)

89. Husband and son read novels and papers. Wife never reads.

90. Wife very fond of reading, and prefers Mrs. Henry Wood.

91. None of the family ever read, which the old grandmother thinks ' a sad pity.'

92. Husband does not care for reading.

93. None of the family ever read at all, except the daily paper. (Wages, £3 per week.)

94. No readers.

95. Fond of reading 'good stories' and papers. (Wages, 3s. 3d. per day.)

96. A great reader, especially travels and biographies.

97. Daughter's husband reads aloud to the family in the evening.

98. Read only the papers. (Wages, 38s. per week.)

99. Reads a great deal—everything he can get hold of. (Wages, 36s. per week.)

100. Do not care to read at all. (Wages, £2 10s. per week.)

101. Mother and son both fond of reading novels.

102. Does not care for reading, except the newspapers. (Wages, 5s.)

103. Great reader. (Wages, 36s. per week.)

104. Husband great reader. Wife never been taught to read.

105. None of the family care in the least for reading.

106. Both fond of reading, especially Mrs. Henry Wood. (Wages, £2 10s. per week.)

107. Not a reading family, but like to hear the news. (Wages, 5s. per day.)

108. Wife 'puts in the days reading and sewing.' (Wages, 26s. per week.)

109. Wife very fond of reading. As a rule the men do not read many books, but they nearly all

[See page 39.

Tapping the Furnace: I. Clearing the Hole.

[See page 39.

Tapping the Furnace: II. Driving the Bar.

[*See page* 42.

Metal-Carriers lifting Pig Iron.

[*See page* 43

The Slag Tip.

read an evening paper. (Wages, 30s. to 45s. per week.)

110. Always reads an evening paper, but does not 'bother with books.'

111. Read evening paper.

112. All very fond of reading, and get books from library every week. (Wages, £2 per week.)

113. Spends Sunday in bed and reads all day. (Wages, £2 per week.)

114. Son reads all his spare time, *Tit-Bits* and *Answers*, and goes in for the competitions, especially those to do with football. (Father's wages, 3s. 6d. per day; son's, 5s. 6d. per day.)

115. Wife can neither read nor write.

116. Not fond of reading. Wife likes reading story-books of every description — 'nice ones.' (Wages, 35s. to £2 per week.)

117. Boys great readers.

118. Neither husband nor wife very fond of reading. He reads *News of the World* and the *Gazette*, and she reads the *Sunday Companion* on Sundays.

119. Apparently fond of reading, as there were several penny books lying about.

120. Cannot read, and his wife no time for it. (Wages, 23s. per week.)

121. Never cared for reading much, and wife cannot read.

122. Husband very fond of reading—anything he can lay hands on.

123. Great reader; novels and newspapers chiefly —'nothing deep.' (Wages, 30s. to £2 per week.)

124. Reads sporting papers. (Wages, 30s. to £2 per week.)

125. Very fond of reading, especially anything about his work.

126. Does not care to read anything but the newspapers.   Wife prefers to sew.

127. Do not read anything.

128. Neither he nor his wife can read, as, though they were respectably brought up, they were sent out to work very young.   (Wages, 25s. 6d. per week.)

129. Both very fond of reading—she *Horniman's Magazine,* he the papers, as he likes to know what is going on in the world. (Wages, £2 10s. per week.)

130. Neither he nor his wife care in the least for reading — she prefers sewing, he plays with the children or sleeps over the fire.

131. Wife a good scholar and likes reading story-books.   Husband can hardly read at all, but listens to his wife reading.   (Wages, 27s. to 30s. per week.)

132. Very fond of reading stories of sea adventures.   (Wages, 30s. per week.)

133. Reads the evening paper, but never books. (Wages, 3s. 9d. per day.)

134. Reads a great deal, but only books on 'religious disbeliefs.'   (Wages varying from 23s. per week.)

135. Not a reading family.   Man reads the papers.

136. Likes to read everything he can find ; nothing

comes amiss, as he is a really good scholar. (Wages, 27s. per week.)

137. Does not care for reading at all. Wife likes reading Mrs. Henry Wood's books. (Wages, 36s. per week.)

138. Both fond of reading, especially Rider Haggard's books. (Wages, 36s. to 39s. per week.)

139. No scholars. Eldest boy tells them 'all the news he hears.' (Wages, 27s. to 29s. per week.)

140. Wife cannot read. Husband reads aloud to her in the evening while she mends.

141. Single man, not fond of reading.

142. Wife very fond of reading histories and the customs and modes of living in other lands.

143. Both very fond of reading 'stories about other people's lives, without any murders in them.' (Wages, 22s. per week.)

144. Single men. Very fond of reading, especially biographies.

145. Son is very fond of reading 'books of adventure' and 'wild escapes.'

146. Reads the newspaper 'without much interest' —never reads a book. (Wages, 4s. 6d. per day.)

147. Does not care for reading. (Wages, 30s. to 36s. per week.)

148. Both are fond of reading nice tales of home-life. Girl-wife appreciates poetry. (Wages, 4s. 9d. per day.)

149. Wife has never learnt to read. Husband reads to her nearly every evening, chiefly religious books.

150. Reads books on his work whenever he can get them, as he is most interested in it.

151. Reads the paper intelligently, and can talk about what he reads. Wife says she is 'not struck on reading at all, and could never see any interest in it.' She thinks her husband reads too much, as he always reads the evening paper. (Wages, £2 per week.)

152. None of the family care for reading. (Wages, £3 per week.)

153. Neither care to read. (Wages, about £2 per week.)

154. Reads the papers.

155. Served in Boer War; reads eagerly all accounts of the war; fond of books of adventure and the lives of explorers. Wife hardly ever reads, not caring for it. (Wages, about £3 per week.)

156. Reads newspapers only. (Wages, £2 per week.)

157. Never reads anything, not even the newspaper. (Wages, 5s. daily.)

158. Cannot read. Wife suffers with her eyes, or would read a good deal.

159. Does not read much, as he is 'too tired' after his work. Wife would read more than she does, as she is very fond of it, but she does not like the sewing to get behindhand. (Wages, £2 and over per week.)

160. Whole family fond of reading novels. (Wages, £2 per week.)

161. Does not care in the least for reading of any

kind. Wife has 'too much to do' to be able to read.

162. Neither he nor his wife can read. (Wages, 23s. per week.)

163. Does not like his wife to read—thinks she might 'get hold of wrong notions.'

164. Both man and wife fond of reading 'all nice tale-books. (Wages, 36s. per week.)

165. Not fond of reading.

166. Only reads daily papers. Wife '*very* poor scholar.' (Wages, 36s. per week.)

167. Great reader, particularly fond of Dickens, and *Chambers's Journal.* Wife thinks reading *most uninteresting ;* ' it always sends her to sleep.'

168. He reads aloud to his wife in the evening Mrs. Henry Wood's books and Dickens. Wife too busy too read. (Wages, 21s. per week.)

169. The second son in this family is 'very intellectual,' and a great and enthusiastic reader of what his mother calls ' such dreadfully hard books.' He is a Greek scholar, and possesses numerous Greek books, besides Ruskin, Browning, Shakespeare, biographies, etc.

170. A great reader, and borrows novels at the club he goes to.

171. Both man and wife fond of reading novels, and he also likes travels.

172. Reads the daily papers assiduously. (Wages, 36s. per week.)

173. Both he and his wife read a good many novels during the winter evenings.

174. Both fond of reading 'good novels—the old-fashioned sort.' (Wages, 36s. per week.)

175. Only reads evening paper.

176. Fond of reading about travels; very loath to lend his new books to the neighbours. (Wages, £3 per week.)

177. Both he and his wife read newspapers and magazines, but 'cannot be bothered with a book.'

178. Both very fond of reading. (Wages, 36s. per week.)

179. None of the family care in the least for reading.

180. Both fond of reading; he reads aloud to her while she makes the children's clothes. (Wages, £4 per week.)

181. Reads paper only.

182. Reads 'good novels' to his wife.

183. Very keen politician, and reads all Parliamentary news.

184. Neither he nor his wife ever read a book. (Wages, 25s. per week.)

185. This woman, at the age of fifty, made a desperate attempt to learn to read, and, being asked what sort of books she would prefer, said, 'Something with a little love and a little murder.'

186. This woman, asked what she liked to read, replied: 'Something that will take one away from oneself.'

187. Husband and wife read daily local paper—nothing else.

188. This family never 'wastes time' over reading

books, but the father and son take an interest in the political news in the papers.

189. Parents fond of reading penny novelettes. Son prefers sporting papers.

190. The husband does not care to read more than the evening paper. Wife cannot read or write, but she gets her husband to read to her all that is going on in the world.

191. A rather delicate young man, fond of reading, especially books about his work; explaining his love for it by the fact that his ' father was the same, and had more brains than he knew what to do with.'

192. Husband reads.

193. Husband 'sits down with a paper or book' after his supper every evening.

194. Reads only newspapers. ' No time ' for books.

195. Young man, very ill, likes the picture papers, and is especially interested in sports, cricket and football.

196. This woman, when asked if she never read, answered she ' was no scholar,' and she was often glad she had never learnt to read, as if she had she might have ' put off '—*i.e.*, wasted—the time, instead of doing needlework.

197. The husband much interested in reading, especially anything about Japan.

198. Husband takes evening paper, reads Dickens and Thackeray, and 'always has a book going.' (Wages, 22s. per week.)

199. Husband does not read books, but reads many of the papers. Takes in football papers and *Answers*.

200. Husband and wife have a Canadian paper sent them every week by relatives, and read every word of it.

The above number (200) is perhaps hardly enough to generalize from, but still it is interesting to see that of that number there are :—

17 women who cannot read.
 8 men who cannot read.
28 houses where no one cares to read.
 8 men who actually dislike reading.
 3 women who actually dislike reading.
 7 women who say they 'have no time for it.'
50 houses where they only read novels.
58 houses where they read the newspapers only.
37 houses where they are 'fond of reading' or 'great readers.'
25 houses where they read books that are absolutely worth reading (this includes the men who read books about their work).

Some of the readers among the above are men of the very keenest intelligence, reading the best books that they can lay hands on, and eagerly availing themselves of the very good Free Library belonging to the town in question. The population of the town is over 100,000 ; the number of borrowers from the Free Library is 4,500, that is, 4½ people out of every 100 take out a book. In many cases, of course, this represents a larger number of readers than borrowers, as books may be passed on to other members of the family; but it does not represent, as may be seen, a great number of the public. Statistics of the mere number of volumes taken out may be misleading, as the same people take out books over and over again. Most of what is taken out of the Free Library is fiction. The Library is used by many of the better class of workmen, but not much by the very poor. It is quite possible that some of these are deterred by the mere ceremonies that have to be gone through to take out a book. A woman who lives in a distant part of the town, whose outer garment is probably a ragged shawl fastened with a pin, may not like going up an imposing flight of stairs, getting a ticket, giving a name, looking through a catalogue, having the book entered, etc.; whereas many of these would read the book if it were actually put into their hands. Women, at any rate, of all classes know how often our actions are governed by our clothes, and how the fact of being unsuitably clad for a given course of conduct may be enough to prevent us from embarking on it. The establishment in recent years of children's libraries connected with the schools is giving admirable results. The children get into the habit of borrowing books and taking them home, and are more likely to frequent other libraries as they grow up.

The people who, for one reason or another, do not use the Free Library, will sometimes be willing to frequent smaller and less imposing centres of improvement. In two small lending libraries connected with ironworks, the one standing actually in the midst of the works, the other in a workmen's club in the town, the reading-rooms are well frequented, and give the impression of a goodly number of readers; but, as a matter of fact, on the works' library list there are about 70 borrowers out of a possible 1,000 or more, and at the workmen's club 60 out of a possible 600. There are here, again, however, more readers than there are borrowers on the list, as each book is handed round and read by several people. These somewhat unambitious libraries have been gradually provided with books, on the principle not so much of directing a course of reading as of providing a course that would be acceptable to the readers. They contain children's books, most of Dickens, most of Scott, some of Miss Yonge, Hall Caine, Bulwer Lytton, a good many of Mrs. Henry Wood, a number of miscellaneous tales of a harmless kind, and also some books that may be grouped together as 'improving,' such as—I quote at random from the catalogue—'A Chapter on Science,' 'Voyages of Columbus,' 'A Dash from Khartoum,' 'The Great Boer War,' 'The Great Invasion of 1813,' some poets, some Shakespeare, and some essays. The latter books are not much taken out; again here, what is chiefly required is fiction. This need not surprise us. The town we are speaking of exists for the iron trade, and is inhabited mainly by workmen engaged in strenuous physical labour ; they are not very likely at the conclusion of a day's work to wish to read anything that involves an effort of attention. Many busy hard-working men in other walks of life read when they come in from their day's work, I believe, just about the same thing, with a difference, as the hard-working man of the artisan class : that is, they read the papers and they read novels, or at any rate something that is purely recreative : and the desirable thing, no doubt, is to have something recreative, more or less interesting, available at the right moment.

It often happens in a workmen's club that a man will take up and read a book that he finds on the table, when he would not even ask to have it got out of a cupboard in the same room. And this also is quite natural. A working-man seeking diversion may be willing to read the things that he finds under his hand, but he may not have purpose and zest enough to take definite steps to procure anything else, let alone the fact that he may not know what to procure, since he has not the opportunities enjoyed by the better off of compiling lists of books from the literary columns of the newspapers.

It will be interesting to observe that of the authors mentioned in the above list the name of Mrs. Henry Wood occurs seven times, Shakespeare twice, Dickens twice, Marie Corelli once, Miss Braddon once, Rider Haggard once. The chief favourite, therefore, as may also be gathered from the books taken out of the libraries, is Mrs. Henry Wood. It is interesting to consider the reason of this. Mrs. Henry Wood has doubtless delighted many of the educated when they were younger, that is, before their experience had shown what we will call the unlikelihood of some of her combinations. What makes her so popular among the working classes is probably, first of all, the admirable compound of the goody and the sensational : the skill in handling which enables her to present her material in the most telling form, and a certain directness and obvious sentiment that they can understand, while at the same time it is just enough above their usual standard of possibilities to give an agreeable sense of stimulus. ' East Lynne ' is perhaps the book whose name one most often hears from men and women both. A poor woman, the widow of a workman, who had gone away to a distant part of the country, and was being supported by the parish, wrote to some one in her former town to say that she thought that if she had ' that beautiful book " East Lynne " it would be a comfort to her.' And on another occasion a workman, wishing to add to the library of a club he frequented, brought a copy of ' East Lynne,' saying it was the book he liked best.

The working-men's wives read less than their husbands. They have no definite intervals of leisure, and not so many of them care to read. Among those who do, most of them, I am told, prefer something about love, with a dash of religion in it. This is the character of most of the penny stories which form the bulk of the literature accessible to them. They like some relief to the greyness of their lives, some suggestion of other possibilities; but for many of them anything that excites laughter goes too far in the other direction, although they are usually ready to laugh at something humorous if read to them more than if they read it themselves. But they generally prefer something emotional and not laughable. More than one would expect of the women between fifty and sixty cannot read: even some of those of forty. And they nearly all of them seem to have a feeling that it is wrong to sit down with a book. If there is anything more practical to do, is there not some truth in this view? Even among the well-to-do this idea persists a great deal more than one would at the first blush admit.

The women who never read during their leisure usually gossip. It is not so much that what they read is beneficial to them as that it keeps them from doing something less beneficial. Also, and this is an important factor, it puts them at a disadvantage with their children, and prevents them, even if they were disposed to do so, which does not happen very often, from exercising any check over the children's reading or taking an interest in it. There are a certain proportion among the women who are book-lovers, who will read anything that comes in their way; and heavy is the responsibility, therefore, of those who should help to provide them with books. I was told by one of the nurses at a Home that a working woman she was nursing had seventeen penny papers, which she used to conceal under her pillow. Other reading was provided for her of a rather better character, some of the better six-penny magazines, which she devoured. She could hardly be taken as a type, as she was of the genus omnivorous reader.

I have looked through a number of the penny stories that the women mostly read. They are irreproachable, and they have the most curious resemblance of plot. In four that I read, one after another, the poor and virtuous young man turned out to be a long-lost son, and became rich and powerful. One thing to be deplored is the very small print of these publications. One can hardly expect for a penny to get a complete novel in pica; but there is no doubt that the very bad print of these books, read often by imperfect light, is largely responsible for the damaged eyesight and headaches among the women, as they grow older. It would be, to my mind, a public benefaction if it were possible to organize some distribution of books on hire, in good print—wholesome, harmless fiction, of the kind that would interest the cottage readers—literally a cir-culating library for the people. A hawker with a barrow might carry these round the poorer streets, and offer them on hire at the cottage doors, as so many things of other kinds, that are absolutely useless to boot, are now carried round and offered. If such books, published at sixpence or even nine-pence, were let out at a penny or twopence a week, according to their size and price, the hawker would probably drive a good trade. The selection of books, of course, would not be left to him, and, at the beginning, he would have to be guaranteed against possible loss. He would certainly find customers ; for, as we have shown, the working-man's wife, as a rule, is ready to hire anything that is offered to her.

The born readers, men and women both, are of course not so dependent upon what is put into their hands: but the great majority are hopelessly depen-dent on it in this class as in any other. It is perhaps worth adding here that on finding what were the results of the inquiry made respecting reading among the workmen, a similar investigation was attempted among people who were better off, and the result of this inquiry among those whom we may call 'drawing-room readers' is curiously instructive. The first fifty who were asked had all during the previous six months been reading the same books. On all their tables there were five or six large biographies, a book of essays, some letters that had attracted attention, one or two novels by personal friends, one or two novels by writers of position: that is, all these educated people had been reading, exactly like the uneducated, the books that came under their hand or that other people had talked to them about. In very few cases had the average reader who had not the temperament of the student, gone beyond this to strike out some line indicated by a particular bent, aptitude, or special course of study.

In the face of this very hand-to-mouth course of reading of the well-to-do, we can hardly wonder that the average working-man and his wife, who do not hear much talk of books or of writers, should not eagerly seek for the masterpieces of literature, or gravitate, as we seem to expect them to do, towards the very best. It seems undeniable that for the great majority of people reading means recreation, not study: it is a pity we have only the one word to designate the two pursuits. And we may well rejoice, and not seek for anything further, if the working-man, and especially the working woman, whose daily outlook is more cramped and cheerless than that of the man, should find in reading fiction a stimulus and change of thought.

CHAPTER VIII

THE WIVES AND DAUGHTERS OF THE IRONWORKERS

THE key to the condition of the workman and his
family, the clue, the reason for the possibilities and
impossibilities of his existence, is the capacity, the
temperament, and, above all, the health of the woman
who manages his house; into her hands, sometimes
strong and capable, often weak and uncertain, the
future of her husband is committed, the burden of
the family life is thrust. It is, no doubt, an obvious
platitude to say this, but it cannot be too often
repeated in these days when we are constantly,
anxiously, unavailingly, trying to prevent the much-
discussed deterioration of the race. The pivot of the
whole situation is the woman, the wife of the workman and the mother of his children. Happily for
this country and its people, there are many of the
working women who carry that burden, in spite of
its enormous weight, with a courage and a competence
alike marvellous to the onlooker. Various observers
who are in constant intercourse with the ironworkers
are of opinion that in quite half the homes visited

such women as these are to be found, women who steer their difficult life in a way which is at once a consolation and an example. Some of us are inclined to think that this estimate may err a little on the side of optimism. It is obviously a question that cannot well be dealt with after the precise and definite fashion of ordinary statistics; nor is it always possible to reproduce in print knowledge obtained in confidential conversation with the working people concerned. Experience brings the conviction that the majority of women, handicapped as they are by their physical condition and drawbacks, have but just bodily strength enough to encounter the life we are describing. The married woman of the working classes has to fulfil single-handed—often on less food, less health, less strength than her more prosperous sisters—the duties that in the houses of the latter are divided among several people, and even then not always accomplished with success. There are many of us who with the assistance of some one else can bear a burden that we should not be able to carry alone without breaking under its weight. I do not wish to make any sensational comparisons for the sake only of effect; it is easy enough to surprise and horrify by the juxtaposition of glaring contrasts. But this is a question with regard to which comparisons must be made and contrasts studied if we would approximate to realizing the task that lies before the wife of the workman. Temperaments, characters, predispositions, aptitudes, are presumably distributed in much the same proportion in every layer of the social stratum ; but achievement and result must depend to a great extent on the surroundings under which these activities are brought into play, and on the more or less excessive demand made upon the energy of the individual. In every class we find the woman who is brisk, confident, energetic, and able to deal with life under almost any aspect ; but I fear women of this kind are in the minority. The majority of women everywhere, probably, are those who have energy enough to go straight ahead if no great demands are made upon their mental or their physical resources, but not otherwise. Most of the prosperous as well as unprosperous women we know are of this kind. There are a thousand of us who can walk along a level road and get to the end of it successfully, for one who can swim a river or scale a cliff which stands in the way. But in the class I am describing, the women who marry do have a great demand made on their resources, mental, moral, and physical, and it is wonderful to see how many of them succeed in keeping abreast of life and making a success of it. Over and over again one goes into one of these cottages in a little back street of the town, or in a wind-swept tract at the works, inhabited perhaps by a husband and wife with several children, living on between 30s. and 40s. a week or less, and one finds it the

picture of a happy home, the husband and wife
devoted to one another and to their children, the
house administered with order and method, the man
made comfortable, the children well turned out. It
is true, as has been said, that, however prosperous,
they are all walking very near the brink; but even
when illness or disaster comes, it often brings out
still more the unselfish devotion to one another with
which it is encountered.

Such was the home of a very competent workman,
Jack D. The household consisted of the husband
and wife and two daughters in their teens. This
man, who had a delicate chest, caught a chill at the
funeral of one of his mates. His wife begged him
next morning not to go to his work. But he insisted
on going all the same, as he did not want to lose a
day's pay. He was not working on the eight-hours
shift; his hours every day were from 6 a.m. to 6 p.m.
It was his habit at night to send his wife and
daughters off to bed an hour before he retired, and
then have a quiet hour's reading by himself, including
a chapter of the Bible. The first thing that made
his wife realize, a day or two after the friend's
funeral, that he was really ill was that instead of
remaining downstairs an hour, he came upstairs
almost as soon as they had left him. The wife then
went and said to her daughter, ' I'm sure father isn't
right, he's not biding downstairs to-night. . . .' And
the daughter, as anxious as she was, watched with

her during the night at the man's bedside. He was
after this off work for several weeks, which meant
that during all that time he earned nothing, and that
what they had to live upon was 15s. a week from two
sick clubs to which he subscribed, and such meagre
savings as they had been able to effect out of 36s. a
week. Besides this a 'gathering,' that is, a collection, had been made at the works for him by his
fellows among themselves, amounting to some £5, contributed with the unfailing generosity always shown
by the men under such circumstances. In this case
the feeling of good-will was increased by the fact that
Jack D. and his wife were both deservedly
popular. During the whole of his illness, in which,
as I have said, no money was coming in, and everything was more difficult, he was tended night and
day by his wife and daughters with a solicitude and
an affection that could not have been surpassed.
Afterwards, when the long weeks of illness were over,
during the man's weary, dragging convalescence,
when the resources of the savings and of the gathering had come to an end and the 15s. from the sick
clubs was all that remained, the house still looked as
clean and spotless, the wife was as uncomplaining,
and the whole household as peaceful as before. And
the visitor who went to that house came away with
the indelible impression of the aspect of the worn and
pale wife, almost at the end of her strength after the
weeks of nursing and privation, and now face to face

with the mortal anxiety for the future, still presenting a front of courage and cheerfulness. It is good to know that such homes exist, and that this is far from being the only one of its kind.

Another very delightful household was that of Peter F., who had a wife and seven children, to whom he was devoted. The mother, a delicate woman, always ill and ailing for some time before and after the birth of her children, succeeded in spite of incessant weakness and ill-health in organizing her house with care and skill. The work of it was divided between herself and the children old enough to take part in it. One child kept the yard clean, another the outhouses, another cleaned the weekday boots, another the Sunday boots, another the fenders, and each one in turn helped with the 'washing up.' These people had £2 5s. a week, and sometimes more, a good income, no doubt, for a man living in a four-roomed cottage, and with no obligation to dress in broadcloth; but still, it is not income only that is needed in order to obtain so satisfactory a result. There is many a man at the works in receipt of as much, whose wife is kept in ignorance of the amount he receives and is made to struggle along on only a proportion of it; and many another, whose wife, even if she receives the whole of his earnings, is quite incapable of administering them to the best advantage.

Another woman, Mrs. M., whose husband has worked for the same employers for thirty years,

governed her house with the skill of a born adminis-
trator. Her two elder daughters were working in
linen-drapers' shops and she did all the housework
herself, arranged day by day with regularity and
method. On Monday she brushed all the Sunday
clothes, folded them up and put them away; on
Tuesday she swept thoroughly upstairs, Wednesday,
she did the week's washing, Thursday she ironed,
Friday she baked and 'black-leaded,' and on Satur-
day she cleaned the lower part of the house, kitchens,
etc., and did most of the cooking for Sunday.
On Sunday all was ready for a day of leisure and
comfort. The result of having a wife of this kind at
the head of the house is that the husband after he
comes home from work, seldom stirs out of the house
again. This man's particular 'job' at the works is
not a very arduous one; he comes home, he says,
quite fresh, and not tired; but having been all day in
the open air and on his feet he does not want to go
out again when he gets home and finds a comfortable
meal. And happily there are many houses where
one hears the same thing. Whenever a man says
that he does not 'care to turn out again' one may
be sure that the wife understands how to make
things comfortable for him. Instances of this kind
might be multiplied: it is well to dwell upon the fact
that they exist. Our attention is so frequently, so
urgently and so rightly called to the dire results
brought about by the ignorance and incompetence of

even the well-meaning among the women, as well as by their ill-health, that it is good to remember that there is another and brighter side that must not be overlooked.

It is of course idle to try to make any hard-and-fast estimate of the proportions that the light and shade of this picture bear to one another; but it is of paramount importance to face the conditions under which the dwellers in the less fortunate homes, and especially the women, lead their lives, and to see what can be done in the way of remedy or of alleviation.

The position of the women of the working classes in Middlesbrough is different from that which they occupy in the other big manufacturing towns, for the reason that what is practically the only large industry in the town, the iron trade, offers absolutely no field for women in any part of it. There are not, as in most manufacturing towns, large factories; there is therefore no organized women's labour. The women have no independent existence of their own. They mostly marry very young; the conditions of the town point to their doing so. It is one of the few in the kingdom in which, from the constant influx of men in search of work, and not of women, the males outnumber the females. In 1901, when the census return was 91,302, the number of males was 46,882, of females, 44,420. The population now reaches nearly 100,000, and the disproportion in the numbers of the sexes has since, from the industrial development of the town and the larger influx of workmen into it, still further increased. Here the girl of the working classes, therefore, usually marries in her teens, or soon after. If she wishes to earn anything before this, when she is growing up, she either goes into a shop or into domestic service. The shops, of course, do offer employment to a certain number, and there are one or two small factories. It is perhaps worth while to show here what are the industrial opportunities open to the daughters of the ironworkers, by giving a list of the women employed in a given year in Middlesbrough by the following industries:

Dress and mantle makers ...	380
Tailors	136
Milliners	110
Hosiery-knitters	53
Sweet-boilers	34
Marine stores	25
Paper-bag makers	15
Beer-bottlers	13
Bottle-washers	10
Salt-packers	9
Upholsterers	7
Lubricating-bag makers... ...	6
Joiners and cabinet makers ...	5
Bookbinders	3
Slipper-makers	3

Mattrass-makers ...	...	...	2
Badging names on glass		...	2
Picture-frame makers	...	...	1
			814

The first four of these, in which far the greater number of women are employed, are of course callings which offer employment to women everywhere, whatever the special conditions of the town they may inhabit.

Besides the possible occupations shown above, a certain number of girls go into the elementary schools as teachers, continuing, therefore, their school-days without a break by passing from scholar to monitor and pupil-teacher. We need not for the moment dwell on these, who are generally the more intelligent; their life and occupation are provided for until such moment as they marry, and have, according to the rule of the Education Board, to give up their work.

The young man of the iron-working class usually has no misgivings about embarking upon matrimony early and without a sufficient income. He marries very young, often because he wants a home of his own. Either he is in his parents' home, where he is of course not the principal person to be considered, and is set on one side perhaps and has to undergo the discomfort and crowding entailed by being one of a family living in a small cottage; or he is a lodger, under much the same conditions. He marries a girl as young as himself, without in most cases either of them having any preconceived idea of what their life is going to be like, the responsibilities it will entail, or how to meet them. During their courtship they go about together in the streets, for the conditions are usually not favourable or agreeable within doors. I knew one mother whose supervision of her daughter took the strange form of saying she would not have the young man in the house, because, she said, she 'did not like such carryings on,' and she therefore let the girl walk about the streets with him often until midnight. It is not a state of things which is conducive to morality. They 'walk out' together, they fall in love, they have their brief romance. Then these two young people, with the equipment, or rather want of it, for the fight of life that I have described, embark on matrimony, and go into housekeeping on perhaps 23s. or 24s. a week, or even less. On the wedding-day the parents of the bride do not generally go to the church: sometimes the mother remains at home to superintend the cooking of the wedding feast; sometimes she is simply too busy to go out, and the wedding is merely an incident in the daily work. The young man and the young woman go off together to church with another man and girl, who fulfil the functions of groomsman and bridesmaid respectively. They either go on foot, or, if they can afford it, in a four-wheeled cab; and after the wedding is over they go back to their own house, and either have a day off and enjoy themselves, or else the man goes straight back to his work and the wife to her new home. It is not difficult to imagine what in many cases are the lines on which the young wife will start it. Everything depends upon her; the husband's steadiness and capacity to earn are not more important than the wife's administration of the earnings. At first perhaps they get on pretty well. They buy their experience, as most young people do, but the price they have to pay for it in this walk of life is out of proportion heavy. The man's wages, which before marriage generally left him a margin after paying his lodging or contributing to his parents' expenses, need careful handling in his own home to make them go far enough for two. The young wife often does not understand how to do it. She does not know much about cooking, she is not skilful at sewing, she does not know how to organize. At first, however, she may be able to encounter life with tolerable success. Then she has a child, and let alone the fact that during the time preceding the birth of the child everything is more difficult to her probably than before, she afterwards, usually long before she ought to try to do any work, begins struggling with her daily duties again, plus the baby this time, whom she generally nurses, and whom she has to look after entirely. And then,

possibly, before this first baby is able to walk, or
when it is just able to do so, while she is still having
to carry it about and look after it incessantly, another
one is coming, or come, the mother herself, perhaps,
being still in her teens. As the time goes on her
energy slowly ebbs, and with it her courage and her
hope. The average woman under these conditions
becomes more and more unable to overtake the
claims of her existence, to make the physical effort
demanded by having to clean her room, to mend her
children's clothes, to cook their meals; and it is not
so very surprising that she should send for something, however indigestible, from outside to eat,
that she should leave the clothes unmended, that she
should leave the floor unswept. Again and again
one sees the experience repeated of a brisk young
workman, with the makings of a good and successful
citizen, hopefully starting his married life, and then
gradually as his children come into the world and
his wife becomes less and less able to cope with
existence, growing more and more accustomed to
look for comfort and enjoyment out of his own home.
His life at the turning-point is turned in the wrong
direction by the wife; not because she is vicious, not
because she is ill-intentioned, but simply from her
want of competence, from her incapacity to deal
with existence, however she may struggle, and
above all from her failing health. And then they
gradually both of them slide and slide until their

downward progress is hard to stop. These early years of marriage are in many cases the most trying of the working woman's life. The outlook may improve again later when some of her children are beginning to grow up, and can help her either in the housework, or by earning something. But the first years are hard indeed, during which one child after another is being born. Under these conditions the woman of average vitality, spirits, and capacity, who under favourable circumstances would have had good health enough and enjoyed her existence, becomes indifferent to everything but the toil of which she is trying to keep abreast.

It is sad to see many a young woman who started as a bright, nice-looking girl, struggling at first during these years against constant physical discomfort of every kind, and sinking at last into a depressed, hopeless acceptance of the conditions round her. One's heart aches at seeing a girl of twenty-four or twenty-five, when she ought to be at her best, most joyous, most hopeful—at the age when the well-to-do girl, in these days apt to marry later, is often still leading a life of amused irresponsibility and enjoyment—already appearing dulled, discouraged, her form almost shapeless, her looks gone, almost inevitably becoming more of a slattern day by day from sheer incapacity to keep up with her work. We are apt to think that since it is for many reasons undesirable that a woman should think of her dress and appearance more than of anything else, it must therefore be well that she should not think of them at all; but it is probably almost equally undesirable that she should get to a state of disregarding them altogether. To arrive at that frame of mind implies a great deal. A struggling wife, going down in the social scale, with an increasing family and diminishing means, finds as her possibilities of keeping up a respectable appearance decrease, that she more and more loses also the inclination to be in respectable surroundings.

But it must be confessed that when the women are untidy in their surroundings and in their persons it is not only from want of means. Many of the girls and young women we are describing appear to think that in their own homes it does not matter in the very least what they look like. To some degree, of course, this cannot be helped; they have obviously very few changes of clothes, and they do not wear their good ones at home; but they might at any rate be tidy there, and above all, have their hair done, and not keep curling-pins in it all day. They do this from morning till night apparently, walking about at every hour of the day with a flat head glistening with metal curling-pins, in order to have the whole effect of the resultant curls concentrated upon an hour, perhaps, in the evening. Indeed, some wear the pins not as a means to an end, but as an end in themselves. One young woman was used to wear them for a week at a time without taking them out, sometimes even for a fortnight, day and night, till they grew quite rusty; not, she said, when remonstrated with, to keep her hair out of the way, but 'because they are what every one else wears,' and it was less trouble to leave them in. On the same principle it was only on great occasions that her house was made tidy; as a general rule, things were lying about everywhere in the most comfortless manner. The woman with the temperament of the slattern is no doubt to be found in every social layer; but when she has some one who can remedy her failings, it does not matter so much. We all know the slattern who lives in a castle or a mansion; she leaves her things about everywhere all day long. In her drawing-room are odds and ends, the string of the parcel she has cut open, a telegram she had the' day before yesterday, an envelope of a letter she had last week, the scissors she has just got out of the case, the book she has taken down to consult. All this sounds trivial, but it makes a deposit which needs to be constantly removed. What would be the condition of that woman's room if she had not many servants to remedy their employer's failings? Indeed, what would be the condition of her clothes, if it were left to her to replenish every missing hook and button? But she is fortunately in a walk of life in which it is some one else's duty, not hers, to see to the seemliness and order which is expected to surround her.

The misfortune of the working women, of the working classes altogether, is that the pressure of necessity, as far as they are concerned, makes certain actions, certain qualities, which are for many people only desirable, into iron necessities, to fail in which is not simply an individual characteristic, it amounts almost to a crime. We shall understand the relations of existence better if we admit this and do not try to deceive ourselves ; if we frankly recognize that as a matter of fact there is regrettably one code of conduct for the rich and another for the poor. Any lapse from self-denial, from temperance, from economy, from civility even, in the latter becomes immediately and conspicuously visible because it at once and swiftly brings about a deplorable result, giving the impression that in the working classes these lapses are more frequent than they are in those that are better off.

There are certain habits, certain personal tendencies, certain rules of conduct, which may be good qualities in one class, and defects in another, and some of these which seem to be of the most trivial description may sometimes have the worst effects. A little indulgence in lying in bed in the morning is not a crime under ordinary conditions to the woman who is well-to-do; it is, indeed, often a merit, if her health makes it desirable that she should rest. But to the workman the fact of having a wife who is willing to get up in the morning or one who is not, may make the difference between his home being attractive to him or the reverse; it may lead to his going to the public-house, gradually being ruined by drink, and leaving his children to starve.

Most of the husbands, it is true, when they are at work on the early 'shift,' which begins, as has been stated, at 6 a.m., do not expect their wives to get up to see them off, their food being generally put ready the night before. This, after all, is quite a reasonable arrangement. If the wife were to get up before the man goes and give him something at 5 a.m., it would mean that she had to get up at 4.30, and she cannot afford not to have as much sleep as she can get: in many cases not enough in any event, especially for the young mothers whose nights are broken by their children. It is not surprising therefore that most of the wives do not get up before the early departure of the husband. But having let that moment pass, some of them remain in bed till half past eight or nine, others till later. The fact that in many cases much of their food is got from outside makes it unnecessary for them to be up in good time, and to have the fire lighted. One woman, Mrs. B.—this certainly was an extreme case—was lying in bed at noon, the children, literally half naked, playing about the kitchen floor and nibbling crusts that were lying about; when asked why she had not got up, she replied that she had nothing to get up for.

A neighbour of Mrs. B.—Mrs. H.—was just up at 10.30 a.m., her baby not yet washed.

At Mrs. M.'s house, at 11.30 a.m., the mother and two children had just come downstairs; one child was sitting on the mother's knee, the other standing beside her; they were only partly dressed. She said they were late down because it was so cold in the mornings, and they had fallen asleep again, after her husband had gone out to his work at 6.30.

In Mrs. G.'s house the wife stays in bed till dinner-time because they sit up so late at night.

Mrs. W. was just up at noon, her hair hanging down in plaits, her children not dressed. All this, as we have said, amounts to a crime in the woman, who unhappily for her has no means of supplementing her own faulty administration by the help of others; and it would be too much to expect, under these conditions, that the man who comes in from his work and finds his house in the state we have described, his meal, such as it is, not ready, his one sitting-room full of ill-tended and unruly children, he should not be irritable and despondent and go to seek solace somewhere else.

I remember a miserable story, the more miserable because it was impossible to see which way it could be remedied, of a respectable, intelligent youth, who married a delicate, shiftless girl. She seemed quite incapable of managing her house from the beginning; she let everything slide, saying that she was not strong enough to do her own housework.   Finally her mother came to live with them in order to help her.   The mother proved to be as untidy and incompetent as the daughter; they were both of them fretful and complaining, they did not get on well together.   Then the man had an accident at the works and was off work for several months, living on the compensation allowed him for it, during which time he had to remain in his wretched home surrounded by dirt and discomfort.   One longed to see him taken care of by some tender capable woman. But the two untidy, nagging womenfolk he lived with seemed unable to look after him or after the house.   They could not pay their way, they got more and more into debt every day, this, too, preying on the man's mind.   His two little children, neglected and underfed, played about on the stones outside, each with only a ragged petticoat on.   The man was altogether so miserable that when he was able to be about again he left his home and tried unsuccessfully to make away with himself.   Later however he came back again and went back to work, but still depressed and melancholy, as well he might be with such an outlook in front of him.   This man, if he had married a different kind of woman, would probably have gone on being happy and prosperous to the end of his days.

In houses where there are no children life no doubt is in some respects easier; the woman has not such a hard time—as to work, at any rate. It makes all the difference between poverty and comfort, not only from having fewer people to feed and clothe, the slender funds therefore going further, but from not having so terrible a handicap to the health and strength of the mother. But in most cases the result of having no children is that the home is not so happy. The majority of the parents, it is needless to say, love their children, in spite of all the trouble and anxiety they entail. Children make an occupation and an object, they are a reason for foresight and prudence, they are a bond between the husband and the wife, and it is nearly always a great regret to both when they have none. In most cases losing children is a deep and lasting regret to the mother, often making an indelible mark on her life and health. One woman had had seven children, three girls and four boys. The four boys had all died in infancy, and the wife, a sad, sickly-looking creature, said she had 'never picked up' since their death. Another woman, Mrs. S., had had twelve children, of whom she lost nine in infancy. They all lived to be about nine months old and then died of fits. The mother had never got over the successive shocks. And many instances might be given in which one has been told that the mother has never recovered from the grief of losing a loved child.

On the other hand, there are some mothers who take it more lightly. One bright, excitable little

Irishwoman said, 'I lost all my children when they were babies, but it was better they should go when they were young, for now I know they are little saints in heaven.'

Another said of her child of six, 'I knew I should lose her: she was too clean to live.' There is a deeply-rooted conviction, by the way, among the women, that for a child to wish to keep clean is a most disquieting symptom. And there probably is something in it, as it is the nervous, sensitive children in all classes, as a rule, who object to being dirty, and the sturdy healthy ones who do not.

One woman, Mrs. P., had fifteen children, among them three sets of twins. She lost eight of them. The woman was not particularly tidy, nor the house well-cared for, but the remaining children had tumbled up somehow and the mother appeared to be easygoing, good-natured and cheery, almost incredibly so after a life filled with alternations of bringing children into the world and then seeing them die. One is too thankful to find out that sometimes this process can be lived through with comparative immunity.

Not infrequently one finds homes where children have been adopted, sometimes even when the parents have children of their own, which is certainly surprising when one considers what the additional burden must be of having another child to tend.

One thriving and affectionate couple—the man was a blacksmith—who had had no children, adopted a child advertised in the local papers, and he became to them as their own.   The man died when the child was about ten, and to the surprise of the neighbours, who had thought that the boy would have been a great solace to the wife, she calmly announced her intention of 'advertising him again' in the paper, did so, and found him another home.   After which she married her lodger.

Sometimes such children are taken in for payment, children whose parents for some reason do not want to retain them in their own keeping.   These in a crowded home run no better chance than the other children, and indeed, the small sum paid for them probably does not cover very much of the expense.

Out of 800 houses at the works in which the number of children was inquired into—

58 women had had no children.
97 ,, ,, ,, families of nine and upwards.
275 ,, ,, lost one or more children.
370 ,, ,, not lost any.

These numbers, although not given over a wide enough field to found exhaustive statistics upon, are instructive as far as they go.   It is certainly somewhat surprising, as well as consoling, to see in the working out that there are more houses in which no children have been lost than the reverse; for one's first impression is that when the question is asked of the mother in one of these cottages, 'How many children have you?' the answer one fears always comes, 'I have had' so many; 'I have buried' so many.

We must remember also that the number of children who die is—in some cases, at any rate, not to say in many—due to the fact that the child's life has been insured; for many a time the parent is acutely conscious that it lessens the burden of life on the whole that instead of there being another child to look after, its place should be empty and some additional funds come in to compensate for its loss. One woman whose child had just died said in reply to condolence that 'it would not have mattered so much in another week, as by then the insurance would have come in.' Another spoke almost flippantly about all her children dying, and said, 'It is better they died, for I had them all insured.' This, perhaps, can hardly be wondered at, when one considers what the constant physical strain must be, in the case of a large family, merely to keep the children alive. To the weak and ailing mother the child is looked upon often as more of a trouble than a joy, and if insured its death is a positive benefit instead of a misfortune. It is allowed to die, therefore, without making much effort to keep it. Nor perhaps can one wonder at the deplorably increasing number of women who

take measures to prevent the child from coming into
the world at all, a practice which is no doubt
spreading in this community.   On the other hand
one must set against these the number of mothers
who struggle devotedly to keep their children alive,
and sometimes appear to do so quite miraculously;
children, it must frankly be recognized, whose sur-
vival is no gain to the country.   These most prolific
of the population are not, by multiplying their descendants and keeping them in the world, doing the
most good.   It is bewildering and paralyzing in these
days when we are lamenting at once the declining
birth-rate and the deterioration of the children who
are brought into the world, to look at some of the
houses which keep up that birth-rate and see what
is the result on the health of the average working
woman and, therefore, on the health of her children.
On one occasion a visitor was sent for to the house
of some people, the D.'s, who already had several
children and who were having a hard and incessant
struggle to meet their daily wants.   The visitor
who, on arrival, heard with something like horror
that the mother, Mrs. D., a few days before had now
had twins (two daughters), was met almost on the
threshold by the mother herself, up and about, white,
almost worn to a shadow, but beaming with exultation, displaying a twin on either arm.   The poor
little children were so small, so frail, they hardly
seemed to be alive.   The mother was looking at

them with a face transfigured with tenderness and rejoicing.   And a neighbour, who was standing by said in a tone of hearty congratulation, 'Aye, Mrs. D. is that proud! . . .'   And then there followed weeks in which the mother made unavailing struggles to keep these two poor little things in the world, trying to nurse them both, herself insufficiently fed, until finally, to her intense grief, at three months old one of them died.   The other child lived, and as time went on and the father earned more, things were better.

Mrs. V., whose husband was in receipt of good wages, earning from 35s. to 40s. a week, and who had started five years before as a bright and sufficiently healthy-looking girl, had four children, among whom no twins, before she had been married four years.   She had become consumptive before the third was born, and after the fourth her condition became alarming.   Three months after its birth she was still alive, but lying hopelessly ill, and cursing, literally with her dying breath, the conditions which had driven her to her death.   'It is not right,' she said desperately, 'it is wicked that a woman should be killed by having children at this rate!'   She had been ailing from the first days of her married life, and had never had an interval in which to recover her strength.

This is the story of many a woman who marries under these conditions, and it may well seem to the anxious onlooker an insoluble problem. We are accustomed to say hopefully, 'When general prosperity is greater and wages are better, all this will alter.' Here was a man, whose ordinary expenses were those of the workman, earning nearly £100 a year. This betokens tolerable prosperity. Instead of the birth-rate, as far as he was concerned, declining he was giving to the nation a child every year. With what result? A household of misery and disease, and four children who, as one looked at them, one felt had better not live to grow up. No doubt, as we have shown in the first portion of this chapter, the description of this home does not stand for all, and there are, mercifully, many others which present a more hopeful aspect; but it is no good shutting our eyes to the fact that there are many and many of this kind in which the prolific mother has a history very like the one recorded above.

It is not surprising to be told by statistics that the mortality of infants in Middlesbrough is very high. In 1904, out of 2,072 deaths in the twelve months among the whole population 650 were those of children under one year, and not including the many deaths of the children who were just a little older. For one month in 1905 the death-rate in this town was higher than that of any town in the kingdom, and it was brought up, according to the report of the medical officer of health, by the amount of infant mortality. This high mortality is attributed by experts to overcrowding, bad atmosphere, bad air, maternal ignorance and negligence, unsuitable feeding, an inadequate or polluted milk-supply. It is estimated that of the over 600 children under twelve months who die in one year in Middlesbrough one-third die from preventable causes, mostly improper feeding. The cause of death in many of the cases is given as premature birth, and is accounted for partly by the mother being physically worn out and unfit for child-bearing owing to the short interval between each birth.

One woman had	6	children in	8	years.
,, ,, ,,	7	,, ,,	10	,,
,, ,, ,,	9	,, ,,	11	,,
,, ,, ,,	11	,, ,,	14	,,
,, ,, ,,	12	,, ,,	15	,,
,, ,, ,,	15	,, ,,	23	,,
,, ,, ,,	17	,, ,,	25	,,

What chance has the welfare, physical and moral, of the children thus rapidly brought into the world by a mother whose strength, owing to imperfect nourishment and unhealthy surroundings, must be steadily declining under this immense strain as time goes on ? The numbers given are heartrending when one thinks under what conditions, and with how little alleviation to the mother's sufferings these children are brought into the world. And the bare numbers are not all; for if all these children were well and strong, it would still be a strain on the mother, but it is worse than that.

One woman had had seventeen children and twelve had died; another fourteen, of whom eight had died. One woman had had ten stillborn children, in addition to which four more were born alive; another seven, who were all stillborn. It is easy to write these words; it is wellnigh impossible for the ordinary reader to call up the true picture of what they really mean. Women among the well-to-do hardly ever have this terrible experience of having one stillborn child after another. To be going to bring a child into the world; to be constantly ill before its birth, as must be the case if it dies during this time; to have either the awful suspicion that the hope is over, or else to go on to the end; to go through all the necessary agony, to bear a dead child, to have the shock of realizing what has happened; then in a few months begin over again facing the terrible possibility, which in course of time comes to pass, and live through the whole dread story once more. And even when the children are born alive, what must that other woman have gone through who lost twelve out of her seventeen children? When one thinks of the agony of anxiety that it means to watch the illness of one's child, even when every resource is available, what must it mean to go through that anxiety with one child after another in succession, and hopelessly see each one of them die? A woman among the well-to-do who should have had seventeen children and lost twelve, would be one marked out as she went about the world for the wonder and compassion of her fellows; but such a destiny is accepted as possible by the working people, and is cruelly frequent among them. And we must also remember that at probably no stage of these unrelatable experiences has the working woman such possible alleviations as would be the case if she were well-to-do. She has during all these months of past and future misery to do all her housework, to do everything for herself, enduring pain when it comes as best she may, and mostly dependent for outside help and support on the friendly offices, happily so often ready, of the neighbours.

One woman who had been married twelve years had brought nine children into the world, of whom four of the last five, very near together, were two sets of twins. One twin of each set died before the older one could walk, as they were sickly, rickety children. If they had all lived she would have had five children unable to walk; as it was, there were three. During the course of these twelve years this woman's husband was ill for seventeen weeks with pneumonia, and again later for nine weeks, during all of which time she nursed him altogether. And this even for a few weeks, let alone for months, is a strain unrealized by those who, even if they have not one or more trained nurses in the house and if they suppose themselves to be nursing 'altogether,' can always command the attendance of servants outside the sick-room, can have any amount of assistance at their beck and call.

When one considers the precautions with which the well-to-do mothers are surrounded before, during, and after the time of child-bearing, one wonders—if these precautions be really necessary—that so many of the poorer women and children survive.

During the time before the birth of the child, which to the mother in easy circumstances is full of happiness and hope albeit of physical discomfort, the life of the working woman is led with increased difficulty and hardship. The woman who is well-to-do makes her health the first consideration; everything in her life turns upon it, allowances are made for her. She rests as much as she chooses, she can relinquish the duties or the disagreeables of her existence as she pleases. If she is unwise in the ordering of her life at this time, and does things that are undesirable, it is for her own pleasure and by her own wish; no one compels her to do them. But the working woman generally goes on with her daily round of toil until the very last minute before the child is born; she has no one who can take the burden of her life off her if she is tired, cross, depressed, unwell. The moment of childbirth among the prosperous is surrounded in these days in every direction by extremes of antiseptic precaution to ensure healthy conditions, the absence of infection, or of poisoning. Gallons of water boiled for several days previously stand ready for use, the room is spotlessly clean and airy, a well-qualified doctor is ready to be summoned, a trained and experienced nurse is in attendance as well. The mother's sufferings are mitigated by chloroform. After her confinement the whole house turns on her comfort and well-being. If she has older children they are in charge of other people. She is fed with the most nourishing and palatable food, she has a sufficiency of quiet and of sleep, at this moment of the highest importance. Again let us ask, if all these precautions are necessary, or even desirable, what are we to think of the conditions of childbirth in the cottages? In some of these, situated actually in the midst of the works, the ground-floor consists of a kitchen, behind the kitchen another room, sometimes kept as a parlour, sometimes used as a bedroom; opening out of the kitchen is a sort of closet, also looking on to the front, in which there is just room for a double bed. The confinement often takes place in this room, unventilated in winter, too hot in summer, because it is next to the kitchen, and with the air charged with all the impurities that the works send forth, constantly blowing in. In houses that have not this extra little room on the ground-floor the mother is often lying in the front one that serves as kitchen and living - room, though sometimes happily she is in the back room in comparative privacy. It is impossible that the house should be kept quiet, it is too accessible to noise from the outside; and it is so small that every sound inside it is heard. On the other side of the river, in the streets of Middlesbrough, the conditions are much the same; there is probably no better air, though perhaps in the little back streets a little more quiet is to be obtained than in the middle of the works. But there is no isolation possible either in the houses of the town or at the works, for the mother of the new-born infant. One woman, who had suffered a great deal when her child was born, was lying, when it was only a few hours old, in one of the tiny rooms described, to which penetrated the sounds and smell from the adjoining works, and her husband who had come in from his work tired out and had thrown himself down to rest before going to change, was lying asleep in his black working clothes on the outside of the bed.

Nor does it seem to occur to the working people that tranquillity of mind is desirable, even where it might perhaps be obtained, and they do not hesitate to discuss any painful or undesirable topics beside the sick bed of the mother. In Mrs. G.'s cottage a child had been born in the morning, a daughter. The mother had been very ill. She had already had two girls, and the husband had announced that if she had another he would leave her and go to America. Three hours after the birth of the child she looked as if she were dying and as though her only chance were to be carefully tended in peace and comfort and kept absolutely quiet. As it was, the door of the little room through the kitchen, in which she was lying, was wide open and also the front door of the kitchen, which gave straight on to the street. It was in the afternoon, when, under more prosperous circumstances, the mother's hours of repose would have been carefully guarded. Two women were standing by her bed discussing the situation quite audibly.

'I doubt he'll leave her,' said one. And the other said, 'He will that, he's bad to do with when he's not pleased.' 'It's having another lass, you see, that has put him about,' said the first. 'It is that,' said the other. 'She'll have a hard time with him.'

The women, except in exceptional cases, where there is some complication, do not have chloroform for their confinements; they have therefore no mitigation of their sufferings. It must be said, however, that they do not as a rule wish to have it. They have a great dread of becoming unconscious; they think they will die under it. It is illegal to administer it without the consent, sometimes refused, of the husband.

For the time of the confinement a woman is commonly engaged to look after the mother : this attendant, whose empiric knowledge does well enough if all is straightforward, generally comes for a week, for five shillings. Since the Act was passed for the Registration of Midwives, it is no longer legal for a woman without a certificate to undertake this. The Midwives Registration Act, although no doubt in course of time it will tend to the pressure of necessity making such urgent demands for trained assistants that there will be a sufficient supply, has for the moment made the condition of the poor at such times harder than before. For the good old motherly woman who used in the majority of cases to be all that was needed, although she sometimes made terrible mistakes, is not inclined to pay a guinea to have herself registered, or to go through the necessary qualifying and examining which would enable her to do so.

At the end of the week the mother is left to deal with the new-born baby as best she may, and the difficulty becomes greater with each successive child, since the older children have to be looked after at the same time. She is usually up before a week is out ; on one occasion a visitor who went on a Monday to ask after a woman who had had a very bad confinement the Friday before, found her up and cooking the dinner. This is the time when irreparable mischief is often done to both mother and child, since in most cases the young mother is absolutely ignorant of what to do with the baby, or rather of what to avoid. When one considers on what a slender thread the life of a very young baby hangs, one cannot but wonder that so many children of these ignorant if well-meaning young mothers survive.

CHAPTER IX

THE WIVES AND DAUGHTERS OF THE IRONWORKERS
—*continued*

How are we going to attack all the ramparts that stand between us and the possibility of enlightening the mothers? We have to reckon not only with their incapacity in most cases to learn, but also with their unwillingness to learn; and perhaps with a still more serious obstacle, the uncertainty of the teachers as to what should be taught, in what way, and at what stage. And even if we assume that the required knowledge could be imparted, many of the mothers probably would not use it. It is always much more trouble to take precautions than to neglect them, and those after all on whom we would impress the necessity of taking those precautions are already overweighted by having their life full of cares of one kind and another, that the more fortunate are able to leave to others. When we are face to face with some disaster occasioned by sheer ignorance, we say miserably and hurriedly, 'Something must be done'— 'The mothers must be taught to know better.' But how is such a thing to be dealt with, or, more important, prevented? and at what stage of the mother's training? As to including the subject in the school curriculum which is so constantly suggested, it must be a matter of experience to most of us that young people are not in the least interested in learning the theory beforehand of domestic arts destined to be of practical use to them when they are older. How often girls of the well-to-do class, who could never be persuaded to take any interest, any care or thought as to the administration of their mother's house, become most assiduous and careful housekeepers when at the head of their own. One sees such a girl in prosperous surroundings, who never would have taken the trouble to learn academically and theoretically how to deal with a young baby, patiently and perseveringly acquiring knowledge on the subject when she has a child of her own; reading books about it, since, to her, books are easily accessible, and bringing to it a mind accustomed to learning. And during this time, since she has presumably a sufficiency of servants, some one else is obviously doing her house-work and cooking, and she, feeling she is fulfilling a duty in being quiet, has leisure to think, to read, to inquire, to prepare herself for the all-important responsibility she is going to assume. By the time her second child comes she has already, from her experience, from the trouble she has taken to learn from books and from people, acquired enough, in all probability, to enable her to train the child through its first years in safety.

Such is the stage at which for the average woman the knowledge is learnt, and this is the way it is acquired; but it is a way in which it is hardly ever possible to the working woman to supplement her deficient knowledge. She has not the means, the time, the mind, the opportunity or the teacher. But let those prosperous women who remember their own ignorance when their first child came into the world, the trembling diffidence with which they relied on the advice of others, on the skilled trained advice always at their command, until they found their own feet, sympathize with the wives of the poor, who with none of these advantages, are still in that same befogged state of anxious bewilderment when their next child comes, and the next, and the next. By the time they have had half a dozen, of whom, to use their own expression, they have probably 'buried' two or three, there is small chance of their improving in their methods or of their handing on to their daughters much worth having either in the way of precept or example.

The time after the birth of the child, when the mother is supposed to be getting her strength up again, is the moment of all others when care is needed both for the mother and the child. The well-to-do woman is nowadays kept flat on her back

for a week after the birth of her child; and kept
in bed nearly three weeks. The working woman
usually gets up three or four days after the child
is born, and in this condition is soon doing her work
and walking about. She probably is nursing the
child while she herself has not sufficient food. In
these days, when so many well-to-do and well-cared-
for young mothers are not able to nurse their children
themselves, however anxious they may be to do so,
since the requisite nourishment is not forthcoming,
it is interesting to see what a large proportion even
of the ill-fed and delicate mothers among the working
people are able to provide that nourishment, though
it may be of a very poor quality. To be sure, doing
so often tells on their health and on that of the
child. This acts and re-acts; the child is ill-
nourished and fretful and a constant strain on the
mother's care. The mother seeing it is not being
nourished enough, tries to remedy this by getting
it other food. The well-to-do young mother who
cares about her baby, weighs it carefully and
solicitously every week, and when she finds—whether
she is nursing it herself or not—that it is not gaining
weight enough, she concludes that the food it is
having is not agreeing with it, or that it requires
more. She is then able with great care to investigate
the various forms of infants' food in the market until
she finds something which does agree with it; children
vary very much, and what is admirable for one child

often does not do for another. But what is the woman in a cottage to do when she sees that the child is not being properly nourished? In most cases she cannot even afford to buy it milk, since condensed milk in tins is what she is in the habit of using: and it is obviously out of the question for her to be able to buy various forms of food as an experiment and cook them carefully afterwards. All that it occurs to her to do when she thinks her child has not enough is to supplement its diet from her own. That diet is obviously not suitable for babies; but it is given to them nevertheless, even when they are only a few months old, anything that may happen to be on the table, bread, potatoes, jam, pickles. One finds child after child suffering during the first two years from sickness and diarrhœa from being given things to eat that it cannot possibly digest, and this seems to be a practice that cannot be prevented. One could multiply instances. In one cottage a child, fifteen months old, ate exactly what the parents did; this included hot meat, pork, mutton or beef, potatoes, cheese, cake. This child nearly died of catarrh of the stomach, and suffered a great deal from diarrhœa and sickness. Another woman had a child of two years old also suffering from diarrhœa and sickness; she too was in the habit of feeding it on the same food as its parents. Then she tried buying a tumblerful of brandy, which was given with a little water at intervals during the

day. In between times the child could be seen
picking up scraps in the road, and one day when
a visitor called it was contentedly chewing a dirty
cauliflower stalk. 'She is such a one,' said her
mother proudly, 'for picking up anything that's
about.'

Mrs. Y. had had three children with whom she was
quite incapable of dealing. She fed them, apparently
from the beginning, on scraps off her own table, and
when they were ailing did nothing. The eldest of
these three died of consumption of the bowels at
thirteen months old. The other two, looking as
though they were only skin and bone, were suffering
from sickness and diarrhœa. She 'had meant to
take them to the doctor' but had not done so.

In house after house one sees children suffering
in this way, miserable little creatures, often playing
out of doors on the pavement with only one or two
garments on, many of them with their little legs
bowed and bent. It is no wonder that these children,
should they survive the first years of infancy, should
fall an easy prey to measles, whooping-cough, any
prevalent diseases of childhood, especially as it is
extremely difficult to teach anything about the risks
of infection to the mothers, who as a rule take
hardly any precautions to prevent a disease from
spreading.

It certainly is not always from actual want of
means that children are fed on unsuitable food,

since the mother is usually ignorant of what food would be suitable even if she could afford to get it. In many cases to be sure, the more or less sensible women acquire a rough-and-ready knowledge of how to deal with their children, but the remainder display an ignorance, a want of instinct, an incapacity to learn by experience, which is quite unbelievable, and displayed in so many different ways, that it is often difficult to guard against it.

Mrs. V., a delicate mother of several children, the youngest a baby of seven months whom she was nursing, a woman on the brink of consumption, was told that her only chance lay in weaning the baby and in being sent away for a change to pure air and healthy surroundings. This was arranged, and a woman she knew engaged to look after the children. But when the day came and the mother arrived at the new home as arranged, it was found that the only weaning she had done was to nurse the child up to the last moment, and simply come away without it, leaving it to be dealt with by strangers as best they might.

Mrs. T., the mother of a child of five months old which was almost reduced to a skeleton from emaciation, was found to be giving the child pieces of anything she was eating herself. A visitor gave her some Revalenta food and exhorted her to give the child nothing else. At the end of a month the child was fat and flourishing. After which, at the next visit, the mother was found feeding it with black-currant jam as before, and triumphantly told the visitor that the baby was so well now that it was again able to 'get' anything that the mother got.

As time goes on, when the child ceases to be in arms, when it is just on its feet and afterwards, it is threatened by a new series of dangers. As it goes roaming and stumbling about at will in a room crammed for an infant with dangerous possibilities, or running out into the streets it is constantly exposed to risks of every kind. People are apt to deplore the fact that these cottage children are not being trained to obedience. It seems to me that it would be rather surprising, given all the conditions, if they were, although it does happen oftener than one would expect that there has been some training of character and that the children are amenable to authority. But it is oftener, and naturally, the reverse.

A well-to-do mother, who lives in a roomy house with a nice nursery upstairs, can, when her child is naughty or disobedient, ring the bell for his nurse to take him away, feeling that at some cost to herself she has done her duty by him. Let the woman of this kind who thinks she has her children a great deal with her downstairs, knowing that she can, and does hand them on to a nurse as soon as she is tired of them, imagine what it must be to look after them altogether, day and night. What is the other woman to do who has no nursery, no nurse, and no one but herself to deal with the child? who is sitting perhaps with a small baby in her arms, when she sees an older child, not much more than a baby itself, running about and getting into mischief, and being what is conventionally called 'naughty,' that is to say touching, and investigating everything that comes within his ken? If he does not do as he is told, the obvious course seems to be to shake him, or cuff him, or else to leave him alone, none of these being good ways of forming character. And when one room, as is commonly the case of the cottage, is used as sitting-room, kitchen, and nursery, all in one, it is impossible that there should not be danger at every turn. A very common form of accident is that children are scalded by drinking out of the spouts of tea-pots or kettles, thinking that they are the vessels out of which they are often fed; or they go to the fire and touch the grate. In Mrs. K.'s cottage there were two small children creeping about the room close to the fire on which there was no fire-guard. The mother agreed with the visitor that it was a risk and said that some day she would get one. She was evidently quite able to afford it, but said that she always forgot it.

The children brought up at the cottages actually in the midst of the works, in the neighbourhood of railway-lines, trucks, hot iron, pieces of slag, etc., are in specially perilous surroundings, and the river, deep and swift at this place, constitutes another danger for those who are constantly playing upon its banks. For the children who live on the south side of the river in the heart of the town, it is probably more healthy to play out of doors in the street than within doors. But on the whole it is miraculous how many of these infants, with all the drawbacks of their lives, do grow up and are tolerably healthy, often without fresh air enough, without food enough, without warmth enough, without sleep enough, for they are constantly playing about outside in the streets until late at night, or they are kept awake in the house by the grown-up people and by the lights.

They never have as much sleep as they ought at night; they are sleepy and tired in the morning when it is time to get up. The mother is very sleepy herself probably. They are hustled out of bed and hustled off to school only just in time, sometimes with a piece of bread in their hands to serve for breakfast. One can imagine what they feel like on arriving, after a night passed generally in a stuffy and crowded room. They rush back again at mid-day to find whatever food can be provided, which they dispatch with a suicidal and miraculous speed. In one household a small appetizing beef-steak pudding was provided for each child, in a separate basin. This to some of us, although one is rejoiced to find parents capable of providing so definite a meal, might not seem the best fare for children of tender years,

especially when, as in this case, it was absolutely
bolted in under five minutes, after which the children
ran out again to play in the streets.   Then when it
was time to return to school they rushed back again
there, still unwashed and, of course, dirtier than
before.   Such washing of them as had been done in
the early morning was probably of a very summary
description.   Here again no doubt the poor mother
is at a disadvantage.   It is much easier to keep
a child clean when every necessary for its well-
appointed bath is brought up by a nursery maid,
than when the mother herself has to get up in time
to make the fire, heat the water, wash the child and
make its food.   Yet many a working woman can, and
does, do this with quite fair results.

A great difficulty, even with the most enlightened
of them, is to induce them to take any care of the
children's teeth.   For a tooth-brush costs money:
even if it were possible to get it for 1d. or 2d., it
would become a considerable item in the household
expenses if one were bought for every member of the
family, and so in this respect the most unfortunate
results are prepared in childhood.   But we must face
the fact that it is probably not on the ground of
expense only that the brushes are not bought.   This
is a subject on which most mothers absolutely refuse
to believe the necessity of taking precautions.   On
one occasion in a country district in the midst of
a prosperous village community, some well-wishers

harangued the school-children on the necessity of care in this respect, others lectured the mothers : brushes were offered at cost price at the village shop. But no one took advantage of the offer, and yet these people were more prosperous and mostly of a better class than the dwellers in the towns.

The years of schooling on the whole are probably the most safeguarded and the most desirable that some of these children are likely to have, even though it means incessant contrasts of temperature, sitting with wet feet, working when they are cold and hungry. This, happily, is not the case with all of them. There are many who come from careful homes, who are under better conditions, and also many with keenly interested minds to whom, with a stimulating teacher, their school-days can be a time of great enjoyment.

Then comes the moment of leaving school, in which, for the children of both sexes, comes the sudden relaxation of disciplined training and occupation at the moment when these are the most needed.

It has been shown in another place what a difficult problem these years present to the boys.

The girl emerges from her schooling half the time knowing very little that is of practical use to her. This is not said at random, but after a long course of observation, in which delightful, bright little girls have been seen emerging from the sixth standard, many of them quite helpless, some anxious to go into a 'place,' proudly assuming its duties and ignoring the very rudiments of what they are required to do : unable to cook, to sew, to lay the fire, to clean the house.   They can, to be sure, 'mind the baby' with ignorant and loving care, for the majority of girls at that age love a baby; but they have not an idea what to do with it in any emergency.   As we have shown, they have mostly but small chance of learning, since the mother cannot teach them what to do ; and when she can the girl is not always capable of learning.   There are some girls who, arrived at this stage, help their mothers in the house most admirably and efficiently.   When they do, it is generally to the credit of the mother, whose example and training have produced the desirable result.   But more often this form of service in her own home does not offer very much interest to the girl concerned ; it is an intermittent, spasmodic servitude, with no hours of recognized leisure, with no direct remuneration, under the authority of one who, commonly over-worked and overdone herself, considers that she has a claim on her daughter's whole time and energy. The girl in her early teens has a large share of the household work and responsibility thrust upon her, at an age when well-to-do children are hardly allowed to think for themselves.   She has come suddenly on leaving the regular hours of school into an intoxicating sense of freedom and leisure.   Many working women complain that when their daughter arrives at this stage they have quite lost control over her, that they cannot keep her within doors. The young girls between fourteen and seventeen, longing for some enjoyment, are constantly running out into the street; they are often out of doors in the evening till eleven o'clock or later with boys who are knocking about in the same aimless sort of way. It is not surprising, in most cases, that the girl should not be under control. The home atmosphere, ever since the mother began bringing children into the world, has probably consisted of a gradually increasing sense of overwork and inability to cope with it, conditions that do not favour a wise home-training. Nor is there as a general rule very much opportunity for training of character at school, the teacher seeing the child in lesson hours only, without much time for anything besides lessons. The great, the much-needed lesson both for old and young, the need for self-control both in word and in deed, is hardly ever deliberately inculcated in theory and therefore hardly ever practised, a lack which may later produce the most unfortunate effects in the intercourse of the community. Unhappily for the girl who has just left school, this rather difficult join, so to speak, in her life, this moment of assuming the new and multiple duties thrust upon her, coincides with her being at an age when she ought to have less work, extra care, extra indulgence. If she belonged to the more prosperous classes, all these conditions would be complied with. Very often at this time she has irritable nerves, she is fanciful, she is anæmic. But it is also unfortunately a moment when the cares of existence press most heavily on the mother—that is, when the elder children are not earning yet, and there are many younger ones to feed and look after. She is herself nervous and unwell and makes no allowance for the older girl, of whom at the critical time of her girlhood, between the ages of thirteen and sixteen, no care is taken, and it is almost impossible that it should be otherwise. She is not told anything about her health; the mother, indeed, is too ignorant to instruct her on the subject; and any nervous disturbance that may take place at this time is treated as a 'wicked temper,' or met by hard words if not by anything more. There is a much greater proportion of girls of this age ailing and delicate in the working class than among the well-to-do. Indeed, the ignorance and carelessness of both mothers and daughters about this particular phase of a woman's existence is one of the fruitful sources of their undermined health in later years. It does not seem to occur to them that care should periodically be taken of a growing girl. She is constantly standing about, at a time when she ought to be avoiding chill and fatigue, in cold, damp yards, nearly always perforce insufficiently clad, insufficiently shod; or is on her knees scrubbing wet floors. Fitful instruction of an empiric kind is now imparted to her during the process of learning to help in such washing, scrubbing, cleaning, and looking after the children as may be considered necessary by her mother. If slatternly conditions prevail in the home they will probably continue to prevail in the girl's own home when the time comes. She may have had some cooking lessons at school where there were convenient materials prepared for her ; her culinary attempts at home are often left off in discouragement. She is likely to remain an indifferent cook and an indifferent needle-woman, if her mother, as is often the case, has been deficient in these branches. She has been accustomed to see a gaping garment fastened with a pin instead of seeing some one with either the energy or the enterprise to look for a button and sew it on, and she is content to do the same. Then after helping in her mother's house for a time, more or less inefficiently, she perhaps makes up her mind that she will go to service. But the difficulty is that the girl who comes out of the kind of house we have been describing, is probably too rough and dirty to be very acceptable as an inmate of a well-to-do house ; and being a servant in the house of a working man whose circumstances just allow him to keep one, sometimes because the wife is lazy and does not try to tackle the work herself, is often such a slavery that it is no wonder that more and more girls refuse to go into service, since this is their conception of it.

As an instance of this, a girl of fifteen went into
service in a workman's house as ' general,' earning
three shillings weekly.   This growing girl, at an age
at which the child of well-to-do parents is probably
being carefully watched, lest study and occupation
should be overdone, actually did all of the cooking
and most of the work of the house she was in, besides
the whole of the washing for the master and mistress,
and family of six.   She had to get up at 4 a.m., and
on ironing days to sit up till past midnight.   She
struggled on in the place as long as she could, as her
father had been off work ill for several weeks, and
her earnings (she was the eldest of six) were absolutely essential.   No wonder that girls doing this
sort of thing should have but a small chance of
health, nor yet that they should marry the moment
that they can in the fond belief that they will have
an easier life.

To go into a shop is perhaps more attractive; but
even here the conditions are hardly ideal.   If this
girl, instead of going into service, had gone into a
draper's, say, where she might have been apprenticed
to the dressmaking department, she would have had
to work for two years for nothing, having had £1 1s.
deposited for her to learn the trade.   She could then
have left the work at any time, forfeiting the deposit.
After the two years were expired, she would have
earned 15s. a week but had no meals.   Or, she might
have gone as an assistant, three months or six

months at the cashier's desk, and the remainder behind the counter, dinner and tea given but no wages : also giving her work free for two years. Then when she has served her time 10s. per week is the most she can hope for ; it will probably be not much more than 5s. unless she becomes a forewoman or forewoman saleswoman, or buyer-in, one of which is kept at each counter. This girl also will probably marry the first moment that she can. In her teens, probably, she will become a mother herself : delicate, perhaps, and still immature, she will bring one child after another into the world, and then the round and the struggle will begin again.

It is not only bringing children into the world that affects the health of the working women. It is an entire delusion to believe that they are as a rule stronger, hardier, healthier, than the well-to-do. There are a great many women among the latter, probably the majority of them, who are not particularly strong, nor the reverse, and who keep themselves in health by what is called 'taking care,' a process easy enough to the leisured, but not so easy to the workers, even if these know how to do it, which in most cases they do not. They are, on the contrary, as a general rule quite extraordinarily ignorant about their health. They have learnt more in late years, those of them who have had the opportunities of going to ambulance classes, lectures on health, and nursing. Some of these lectures, given by a trained nurse to the wives of the workmen, produced a most admirable result, both in the actual tending of illness and also in the prevention of it. But those who are likely to attend such lectures are nearly always the more respectable among them, therefore those who are likely to be more enlightened already. The miserably poor, the badly clad, those whose aspect shows that they are the people above all others who ought to learn how to take care of their health when it is possible to them, will probably be the last to attend any assemblage where they will be likely to receive instruction.

This is not the place to insist upon or to tabulate certain elementary everyday facts concerning health, of which most of the women are quite amazingly ignorant. As to diet or digestion most of them know nothing, absolutely nothing. They have not even arrived at the stage of knowing that there is anything to know. But, after all, if they had, and if they knew something of the various effects on the internal economy of different sorts of food, they would hardly be able to act upon the knowledge, since most of them are unhappily in the position of having to eat what they are able to get, and not that of being able to get what they can eat. Some of them know how to cook in a rough-and-ready way, and some have learnt more than this; but most of those who have had the opportunity of attending cooking classes or receiving any systematic instruction, have learnt under favourable conditions, with suitable utensils, cooking ranges, and materials provided for them. When they go back into their homes to try to cook there, and have hardly anything that is required for the purpose, it is not so easy to obtain the desired result.

It would be an inestimable boon if there were some arrangement by which an instructor would go round, actually into the cottages, and cook there with the appliances available, pointing out where these were defective, but at the same time doing the best that could be done with the resources at hand. Also, if a wider range of cheap and nourishing foods could be suggested to the women, and if they could be shown how to prepare vegetables as well as meat in some appetizing and nourishing way. But, in the meantime, the diet consists of large chunks of often indigestible forms of meat, washed down at every meal by the eternal tea, and dispatched to boot with hardly any mastication. The condition of the teeth of the working women is a constant source of ill-health, or, at any rate, of discomfort, from childhood upwards, as we have shown. It is no exaggeration to say that one hardly ever sees one of them with good teeth; they cannot take care of them, or if they did, it would be by making a very special effort, guided by self-denial and deliberate intention. And so nearly all the women one sees have either bad teeth, or false teeth, or great gaps in their mouths.

The details of daily care by which well-to-do

women can help to keep themselves in a state of
efficiency, sound almost too trivial to be worth
regarding ; nevertheless, when it means that their
vital energy and spirits are thereby kept up, these
details explain more than we should be at first
willing to admit.

Everyone knows how after a bad night, for instance,
or even after a good one, what an extraordinary
effect a bath and all the accompanying processes of
of rubbing, etc., produce in the way of reviving and
stimulating. Short of this, the mere fact of a change
of clothes, of taking down the hair and brushing it
and putting it up again, the possibility of changing
shoes and stockings when the feet are cold or damp,
of resting, of sleeping during the day, and starting with a fresh supply of energy; all these make
more difference than the heroic might be willing to
admit.   They are nearly all of them luxuries beyond
the reach of the worker.   Neither do the working
women have, as a rule, the exhilaration of brisk
recreative exercise in the open air as an alternative
to quiescence : the enjoyment of beneficial movement
taken at the moment it is needed.   They have, to be
sure, a great amount of compulsory, unregulated
exercise and movement to excess, in accomplishing
the housework and washing, often leaving great
physical fatigue in its train, and the incessant standing is apt to tell upon their health in various ways.
Most of them, it is true, stand more than they need.

A great deal of it is, of course, inevitable ; but even in their moments of leisure, instead of either sitting down, or, if they wish to be in the air, walking, the women prefer to stand at their street door to talk. One of them who was in the habit of standing leaning against her door-post a great part of the day—to the great detriment, by the way, of her housework—complained that she was ‘ that tired she was fit to fall.’ This was in reply to a visitor who had been enlarging on the value of walking and open-air exercise, and had told the women that when they had ten minutes to spare while the children were away at school, it would be much better, instead of standing at the door, to walk briskly to the end of the street and back. But they all scouted the idea of walking as being too fatiguing. Yet the eternal dragging standing is to most women much more exhausting than walking, since there is no alternative play of muscles and no relief.

But it is, perhaps, without due consideration that we are apt to say that this standing and talking at the door ought to be eliminated from the lives of the working women. Social intercourse in some form or another is presumably desirable for most human beings, and in any class of society the conditions under which it is indulged in, whether in streets or in drawing-rooms, often entail a good deal of standing. There are times when it is almost essential to unburden the soul and compare experiences with

some one else, and many of us may entirely sympathize with the woman who, at such a moment,
stands at her street door, looking up and down the
street, commanding a view of any possible incident
of interest, and ready to catch the eye and ear of an
acquaintance.   But there is no doubt that, for some
unexplained reason, there is a pressure of opinion
against the gregarious, among those of the workers
who think themselves 'superior.'   There is an
implied praise of the contrary attitude, a conscious-
ness of virtue in the woman who says, 'I make no
neighbours,' or, 'I keep myself to myself,' which
constitutes, no doubt, an indictment against the
general influence of the talk at the street door.

But as a mere question of health, and not of
character, there can be no doubt that the incessant
standing in season and out of season is undesirable.
It tends to produce the 'bad leg,' so prevalent a
complaint among the working women, arising usually
in the first instance from varicose veins, and becoming a sore which little can be done to alleviate, often
made worse by dirt, neglect, ignorance, and the want
of accessible remedies.

The district nurses who go about amongst the
working women and attend them in their cottages
bring them a great deal of enlightenment—and it is
needed—not only as to the treatment of definite
illnesses, but as to general care and cleanliness.
There are, of course, among the women many who

are born housewives and nurses, as in any other class of the community, and who tend the sick with devotion and skill. Nothing is more wonderful than the kindness of the poor to one another in sickness. They give at great personal sacrifice such help, such services, and such devotion as the well-to-do, able to render help vicariously, are hardly ever called upon to give in person. One sees a careworn woman, who will find it hard to recover any of her rest during the day, willing to sit up all night with a neighbour. Sometimes several of them will practically share the care of an invalid who has no one else to look after her, going in and out, giving up time, energy, that they can ill spare, often giving, also, material help in the shape of food as well. Sometimes those who can cook will make a pudding, or the ever-needed tea.

Many of the working women, it must be admitted, however well-intentioned, do not know much about nursing, and, besides, are not under conditions in which they can do the best for the patient. On one occasion one of the workmen was lying ill of pneumonia in the ground-floor back room of a small cottage, with the door open into the yard, and the dust and smoke constantly blowing in, covering the bed-clothes and everything else in the room. The nurse suggested that the bed-clothes and other linen should be washed. The next day all the linen in the house had been zealously washed, and the wet sheets and towels were hanging to dry on the bed on which the sick man lay, and all about the room round him. And yet what are people to do who live in a small back street of the town, when the day on which the linen must and ought to be washed is rainy, and the only room in the house with a good fire in it is the one in which a sick man is lying? It is not so easy under these conditions and not only as regards washing to comply invariably with the laws of hygiene, even when some of these have been mastered in theory.

The question of washing is always a difficulty in these cottages. There is no public laundry in the town, and the women have to do their washing and, still more difficult, their drying, as best they can. When it is not raining, they hang the linen, if they have no backyard, across the little narrow street, or rather alley, which runs between the two rows of back doors, where it forms a series of damp flapping barriers in the way of progress along the street. The carters who sometimes drive along these back streets have assumed a sort of prescriptive right to get the hanging obstacle out of the way in any fashion they choose, either flicking it up over the clothes-line, or even twitching it off and throwing it into the road; and the women, with some objurgations, accept this course of action as being what they have to expect. If it has been thrown on the ground, they pick it up and wash it over again. The fact that so much of the washing and drying is necessarily done inside these tiny houses accounts for many of the forms of cold to which the inmates are so constantly subject. A cold, a chill, or some other disturbance of health that would have been nothing if the sufferer had been in prosperous surroundings, is, as in the case of the men also, perforce neglected till too late in the compulsory daily round.

Many of the women suffer increasingly from their eyes as the years go on. They are not able to procure glasses, sometimes it does not occur to them to do so if they could; the strain and discomfort increase until constant headaches, if nothing worse, are set up, the failure of eyesight besides affecting all the possibilities of their daily life.

Add to this that the time of life approaching middle age, when women's health is often a difficulty, is always, whether they are affected by it or not, looked upon by the working women as a time when it is inevitable that they should have a great mental and physical disturbance. They encourage themselves to look for it; they make no attempt at self-control, they eagerly watch for symptoms; they are ready to refer any and every passing disturbance to that crisis, either impending or actually arrived. Indeed, it is often conveniently called into account to explain symptoms due to irritability, to fatigue, to indigestion, to any of the snares that lie in wait for those who are not accustomed to self-control either of their tempers or their appetites. This attitude of mind in a woman just reaching middle age, the time which, perhaps, is not the most cheerful moment for any woman, is bound to influence her outlook on life.

That outlook, no doubt, must depend a great deal upon her physical condition. But though health is a potent factor in the woman's general attitude, it is not the only one. In the town we are discussing the women on the whole seem to be curiously devoid of public spirit or interest in outside affairs. I do not know how much this is so in other places, but certainly among the ironworkers one is struck by it over and over again. There are, to be sure, among them women interested in their husbands' work and in the outside world, but there are also many who simply go on from day to day without taking any part in the wider life outside them, without being in the least interested in any public question. At the most acute moment of the Free Trade discussion, for example, in which most of the workmen—not all —took one side or the other, the women, almost without exception, seemed quite indifferent. When they were asked what their husbands' views were, the majority had not an idea, seemed not to have heard the question at issue mentioned, and had no views of their own upon it.

I cannot say what the attitude of the working woman in other countries is towards her husband's work, but there is no doubt that, for the most part, the wife of the English workman is quite detached in interest from it. Many of them don't know when asked what the man's 'job' is at the works; some know what the actual name of his calling is but do not know what it implies. This last limitation, after all, is, perhaps, not so very surprising. There is probably many a wife of the professional class, who knows what her husband's career is, but does not know the details of daily or hourly occupation of which it consists.

One woman, Mrs. Z., whose husband had one of the most arduous 'jobs' at the works, was a typical and extreme case of want of interest in her surroundings, and, indeed, in her existence. She was born in the next street to the mews she now lives in, and this was almost all she had seen of the town. From year's end to year's end she hardly ever went out of the house, excepting to shop as near her home as convenient. It did not occur to her to go out for air and exercise; she had never been down to the river, or across the ferry to see where her husband worked.

Many of the women get into a frame of mind in which they simply accept, in a sort of inertia, what life brings them, as though they had given up the idea, if they ever had it, of the possibility of making anything better of it.

Mrs. N., for instance, a woman who had been bright and cheerful in her girlhood, with a temperament not very buoyant, perhaps, but not the reverse,

became gradually swamped and crushed by the
necessary and inevitable work that fell to her lot.
She had had six children in rapid succession, in
itself enough to affect any woman's health, especially one whose condition does not admit of her
being careful after the children have been born.
Two children had died, which meant that she had
had the additional strain of anxiety, grief, nervous
shock. Her life, she said, seemed to be one ceaseless,
relentless round, in which—no doubt from her incapacity to arrange it, but that, after all, was part of
her misfortune—there was absolutely no time for
leisure. She got up early in the morning, soon after
her husband had started for his work, did some
cleaning, lighted the fire, cooked the breakfast, got
the children up, washed them, fed them, sent them
off to school, did the house, tidied, cooked the dinner—
either for the husband to have at home or to send to
him—mended and washed, got the tea ready, and no
sooner had she 'got cleared away and washed up'
than bedtime for the children began. Then the husband's supper and food had to be put ready for him
to take that night, if he was working on the 10 p.m.
shift, or next morning, if on the 6 a.m. shift. Then
to bed. And the next morning the round began
again, with never one moment, so she said, to rest
or sit down; no change of thought, no relief. The
outlook of every day to this woman and others like
her may truly present itself as discouraging to one

who has not a cheerful, courageous disposition to begin with. A great deal, of course, depends on the initial bias of temperament. The women of this kind ought to have possibilities of entertainment offered to them which would give them change of ideas and bring some relief to the monotony of their lives. The eagerness with which any opportunity for such change is welcomed is an evidence of its necessity. They will all rush to their doors to see any incident that may be attracting attention in the street—a quarrel, or, still better, a fight—indeed, without anything so exciting, the mere passage of a stranger walking down the street will bring every one of them out on to the doorstep to look after him, often to the great discomfiture of the unwary passer-by. A wedding or a funeral they will, of course, go any distance to see, and even the departure on police-court days of the prisoners going by train to the county gaol is a great centre of interest and excitement, and men, women, and children stand in the station-yard round the prison van.

Many of the women, as is shown in another chapter, do not read at all, and do not, therefore, use that means of diverting their thoughts and supplementing and enlarging their limited experience. The result is that they have minds curiously devoid, so to speak, of points of comparison. This, again, tends to their remaining inarticulate. Indeed, one is apt to be misled and to make a false generalization by the fact that the people who are articulate and fluent, and can translate themselves freely into words, are mostly exceptional among the surroundings we are describing. Every now and again a woman of this type is found who has built up some conception of what she thinks life ought to be, and describes it to the listener.

On one occasion a verbal message was sent by Mrs. X., one of the working women, to a visitor, to know whether it was ' right to leave a bad husband,' the messenger waiting for an answer, which would presumably be accepted as final. The visitor, fearing the responsibility of giving an answer in the dark, went to investigate. The woman had had no children : that, perhaps, was the root of the matter. For although a number of children may be a dragging care, the absence of them tells heavily on the life of the wife and on the relations between her and her husband. She was a south-country woman, readjusting herself with difficulty to the conditions of the north. She described her life, which was grim enough as she told it, saying she was chained to a man who was a brute, who had no ideas and no heart, and that her life was spent in slaving for him without getting any thanks. He had killed her soul and he would kill her body as well.

She then described her day, getting up in the morning by 4.30, when her husband was on the early shift, to give him something to eat before he

started, which not many women do ; then seeing him
go off grumbling, cursing, without a word of thanks.
Then she cleaned the house—it certainly was spotless—washed, mended, etc. Then it was time to
get another meal ready. Sometimes he would come
in early, and curse if it were not ready. At other
times she waited on, keeping it hot, knowing what
had happened to delay him, and then he would
come in half drunk, and after the meal go to bed
and sleep like a log, while she cleaned and mended
the clothes he came back in. Utterly worn out at
night she went to bed, to begin the dreary round
again next day. Added to this, when he was drunk he
kicked her and beat her. Her one idea seemed to
be to get away from him and live somewhere else, a
somewhat difficult thing to accomplish. However,
when the moment came that it might have been
possible for her to do so, he having found a job in
another town and she having, in the meantime, had
the offer of some work on her own account where
she was, she decided that she preferred not to leave
him, and they went off together to the new place.

Wife-beating is not so entirely a thing of the past
as some of us would like to think. Another woman
whose husband was not a model of the conjugal
virtues said that, at any rate, she was thankful that
he did not beat her. Another said with pride, when
relating the virtues of hers, that he had ' never so
much as laid a finger ' on her, which shows that it

seems always to be considered as a possibility. This
attitude of the wife, which is more common than
many of us like to admit, is not so much forbearance
as a sort of dogged acceptance of the matrimonial
relation with its rough as well as its smooth side;
and it is small wonder that in many cases the
woman, on her side, without any attempt at concilia-
tion, should make undue use of her chief weapon,
her tongue. It must be conceded that most of the
women have not the slightest idea of self-control, of
not saying the thing that comes into their head,
either to their husbands or to one another, no
matter what it may be. And in moments of annoy-
ance it is often something which is very undesirable
to say; the only hope is that it should not occur to
them to say it. There are, no doubt, exceptions to
this rule, and it may be said that the marriage relation really turns out surprisingly well on the whole,
considering the haphazard way in which it is often
drifted or rushed into. As has been said, the workmen of a marriageable age in Middlesbrough are in
excess of the number of women available for them,
and few girls or young widows remain unmarried.
Indeed, one is sometimes surprised at finding marriage
considered a possibility much later than many of us
would be apt to consider it so. An elderly woman,
the widow of an ironworker, came to a friend to ask
her advice about marrying again. She did not know
the name of her would-be intended; he was the man

who brought round the potato-sack, from whom she bought potatoes to retail to other people. She finally decided to marry him, without knowing in the least what his circumstances were or what his family was. He had a grown-up daughter who kept house for him, and other grown-up children, who objected to the marriage. It did not prove to be a success.

Another case was that of a widow of seventy-nine who had taken in an old man as a lodger. She had supported herself entirely until that age. He had asked her several times to marry him, but as long as she could earn she would not. At last, when she was seventy-nine and he was eighty-one, finding that she could support herself no longer and that he was still earning 35s. a week, she married him, and they lived happily for two years, at the end of which time he died and she went to live with a married son, after which she said sadly that she felt, for the first time in her life, as if she were in a dependent position.

There is a good deal of immorality among the ironworkers, but perhaps not more than in other communities living under the same conditions. As we have shown, the conditions of the lives of the young people before marriage are apt to conduce to it; also, perhaps, the fact that the only way of housing the many single men working in the town is that they should be taken in as lodgers, often into households already cramped for room. There is

not, on the whole, a strong enough pressure of opinion
in the community against lapses from what is commonly called morality to be very efficacious as a
means of prevention. A visitor going to a cottage
one day, found the old woman to whom it belonged
with an infant on her knee, whom she was tenderly
nursing. This woman had two daughters, one of
whom was married, the other not. 'Your grand-
child?' said the visitor. 'Yes,' she replied proudly.
'F.'s child, I suppose?' naming the married daughter.
'No,' the woman said defiantly, 'it's the other girl's.'
And she added, on the defensive against any possible
criticism or disapproval, 'A good job too, she's got
the child and she's not got the man; he was worth-
less, and she is well rid of him.' And this represents
a not infrequent attitude among the people described.
One thing that becomes more and more impressed
upon one's mind after observing many regular and
irregular homes, is that the received panacea of
insisting upon marriage is not always the best for
dealing with the irregular tie, either in a case such as
that recorded above, or in a more permanent home,
where the man and woman have set up house
together, have had children, and things seem to be
satisfactory. In such cases, the fact that the woman
is free to go if she chooses gives her a hold over the
man, and if he ill-treats her she simply threatens to
leave him. I remember one such story in which
a well-meaning visitor had urged a couple to marry

who had lived together for some time contentedly enough and had more than one child : and the woman told her afterwards it was one of the worst days' work she had ever done in her life. As long as they were not married, she said, the man had never dared to go too far, and she could keep him steady by threatening to leave him, but once they were married she had no hold on him, he did not care what he did or how he behaved to her, because he knew she could not get away ; he had gone utterly to the bad, and her life was now miserable. This was only the result of the visitor's well-intentioned advice.

Problems of individual human destiny are not always to be solved, like problems of arithmetic, by received rules. It may well bring to would-be reformers a feeling of bewildered discouragement to find—as probably happens to all of us on occasion—that when they have succeeded in compassing the end recognized as desirable, they have also, by the very accomplishing of their purpose, brought about an undesirable result in another direction. It is a risk that is ever dogging the steps of the enthusiast. We attempt to deal with two urgent problems at the same time, and the solution of one may be incompatible with the solution of the other. I was told the other day a story which, if it had not been well attested, I should have declined to believe. A young married woman was living with her husband and two children on 19s. a week, out of which they

paid 5s. 9d. for their house. This means that the
four of them had 13s. 3d. a week to live upon:
13s. 3d. with which to pay for food, coals, clothes,
absolutely everything, and not one penny more.
And somehow this woman actually managed to do it.
She had so far pawned nothing; her surroundings,
her person, her children were all spotlessly clean.
The eldest child, a girl of six, had been very ill, and
had gone to the Infirmary. She was then declared
to be consumptive, and the mother had been told to
take her away. A young professional man, zealous
for the welfare of the nation, and anxious about the
decline of the birth-rate, seeing one day the tidy
house and the respectable woman at the head of it,
and wishing, no doubt, that it should be people of
this admirable kind who should multiply their
descendants, asked the woman how many children
she had. 'Two,' she replied. 'And how long have
you been married?' 'Six years.' 'Only two children
in six years!' he said. '*You ought to be ashamed of
yourself.*' And he passed on. The woman, up to
that moment conscious only of her daily, hourly
struggle to make the cruelly slender means go far
enough, was left after this amazing interview petrified,
sorrowful, indignant, wondering whether, indeed, she
had cause to feel wicked and ashamed, and whether
she would have been more deserving if she had had
two more children—that is, six people instead of four
—to provide for out of the 13s. 3d. per week left after

the rent had been paid. We shall most of us feel inclined to forgive her for not having done so.

We are gloomily told that the birth-rate is declining. But in many a case, unless the birth-rate declines in his particular house, the workman has not enough to live upon. That is to say, he may have literally enough to keep alive upon, but not enough to prevent him from 'deteriorating.' He needs a good deal more than enough to live upon if he is to be at his best. Then we try to cheer ourselves by laying hold of an imaginary sheet-anchor, and saying, 'He has enough to live upon if his money were handled to the best advantage.' But in most cases his money is not handled to the best advantage, either by himself or by his wife. As we know, he very often does not give her all he earns, wasting, gambling, drinking away the balance that might make the difference between comfort and penury. But even supposing that he gives his wife all he earns, and that the duty of laying it out 'to the best advantage' rests absolutely with her, there are only a certain proportion of the working men's wives who seem to be able to succeed in doing this. The majority of them among the ironworkers, at any rate, do not appear to have this talent, any more than the majority of them have musical talent, or any special development of force or endowment in one direction more than in another. There are only a certain proportion of women in any class who have that particular gift of administering funds rightly, and the number in the working classes is still further diminished by the fact that the mainspring of the woman's capacity, her health, is unduly strained by the incessant demands upon it we have described. Moreover, it is obviously the woman who most needs good management and thrift who will be the most handicapped in exercising them, since the greatest need for these qualities lies with those who have many children, and she who has many children will be, as a rule, less energetic, and therefore less competent.

There are, to be sure, not a few among the working women who, from force of character, helped more or less by soundness of constitution, are able to ' translate the stubbornness of fortune into so quiet and so sweet a style ' that the very thought of them comes as a rebuke to pessimistic generalizations. But there are many others who are not so fortunate. It is no good postulating imaginary conditions in discussing whether we can achieve a desired result ; we have to deal with human nature as it is, and not as it ought to be : and my conviction is that there will never be more than a certain proportion of women who can carry the immense burden allotted to the workingwoman by the conditions of to-day.

CHAPTER X

DRINK, BETTING, AND GAMBLING

As long as the public-houses are practically the centres of the social life of the workmen, as long as they are the most accessible places in which he can spend his leisure, it is difficult to see how drinking is likely to be lessened to any great extent. It is consoling, however, to learn that the amount of drinking in public-houses has appreciably diminished since the legislation which made the publican as well as the drinker responsible for the drunkenness in his establishment. But the most pernicious drinking is not, we are told, that which takes place in the public-house, but that which goes on constantly in the home.

It is not easy to give absolute statistics as to the proportion of men at the works who drink more or less, but it is probably accurate to say that more than half of them do it enough to affect their health and their circumstances. That a man should drink occasionally does not prevent him from being a capable workman, and he is not permanently discharged because he has been seen intoxicated at the works; but it would be too great a risk to keep him there in that condition, and he is fined and sent home till he is sober. It is hardly necessary to dwell upon or to describe the effect of drink on the households, unhappily too numerous, in which it prevails; such descriptions in these days are multiplied everywhere, and are familiar to those who have not the opportunity of witnessing for themselves the terrible results of this tendency. Time after time one finds a household ruined, children growing up half-fed, the wife worn to a shadow, because the husband is drinking away half his money and his family are struggling to live on the rest. I have seen a woman absolutely wasting away with the effort to provide for herself and her children on the £1 a week given her by her husband out of his wages, he himself spending nearly that amount, which would almost have doubled the household's scanty resources, on drink.

A visitor went one day into a poor cottage, where a careworn woman was dusting. The house, though poverty-stricken and almost bare, was clean, and looked as if it were cared for; the children looked pinched and half-fed. Then the woman became aware of an unsteady, heavy step as her husband approached, and she said in a terrified whisper: ' It is the boss coming. You had better go away and come back another time.' And one look at the man as he came unsteadily in told the story of what his wife's daily experience was, her daily struggle to make the pitiful allowance which he gave her, after he had spent the rest of his wages in drink, go far enough.

This story is repeated in house after house. When the woman drinks, as many of them do, the result on the household is probably even worse.

There was a young couple, the M.'s, whose story was a miserable one. The wife, a fretful, delicate, rather silly woman, had begun drinking very soon after their marriage. The husband, a respectable, upright, domestic sort of man, a good citizen and a good workman, gradually found that the money was being frittered away and his home neglected. Finally, depressed and wretched, he took to going out of his uncomfortable home and remaining at the public-house, and began drinking himself. He said he did not care for it, but he liked ' forgetting about things,' and he took absolutely the only course open to him which gave him at once change of scene and material warmth and comfort.

The children in such houses as these are bound to grow up familiar with the sight of drunkenness, and to take it for granted as one of the conditions of daily life.

It is as well to try to realize, when we are considering whether anything efficacious can be done to remedy this state of things, what are the forces arrayed on the one side and on the other in this all-important encounter, and one's heart sinks at finding the overwhelming odds that there appear to be in favour of drink winning the day. The well-meant and untiring efforts of those who are trying to stop intemperance are as nothing in comparison with the efforts of those who are actually trying to further it. As a mere question of numbers, the latter are in an immense majority, and the propaganda they carry on is most potently reinforced by the tendency of those to whom it addresses itself. The latter are more than ready to listen to the one side; they cannot listen to the other.

Among the work-people in question drink takes the place of the pleasures of the table to those who can enjoy them in no other form; it takes the place of temporary excitement and amusement to those who have few other opportunities of either; it is in most cases the inevitable concomitant of the places where social intercourse is to be enjoyed. Unceasing opportunities are offered by the number of public-houses scattered about the town to the men who want to drink, and to the men who do not. In the workmen's quarters there is not a house that has not a public-house within a few hundred yards. There are, of course, a certain number among the workmen, and also among their wives, who have a predisposition towards drink, an inherited tendency, a special bent or craving; there are a still larger number, probably, who have drifted into it simply from circumstances. Opportunity makes the drunkard. The craving for drink on the part of the men is in many cases enhanced by the miserable unpalatable food, often cold and half cooked, that the workman finds in his home, sending him out to seek elsewhere something with warmth and cheer in it. The disastrous habit in the district of ratifying a bargain by drinking may start those on the downward path who never drank before. The stevedores, or stowers, whose business is loading and unloading the vessels that carry the iron to and fro, are practically obliged by public opinion, when the work is finally accomplished, to 'stand drinks round' with all those who have been employed in it. A very decent woman told a visitor that her husband had never drunk in his life until, at first to her delight, he was promoted to be stevedore; then the custom above-named was the ruin of him. The pressure of opinion was too strong for him to withstand it, and, having once begun, he took to going out of his home and drinking regularly.

There are in the town many public-houses which allow their customers to run up bills to a certain amount. This practice, of course, still further facilitates the temptation, as the man with no money in his pocket can call for drink all the same. One man whose wages were 27s. a week, owed 15s. to a public-house. This running account is called 'strap.' It must be very difficult on a cold day to pass one of these well-lighted and well-warmed houses without going in, especially on the way back from a long and tiring day. In the morning, in spite of the fact that alluring placards bearing the words, ' Hot Milk and Rum,' an additional attraction, are displayed in the public-houses as the workman passes, not a great proportion of the men stop to go in on their way to their work. But on the way home it is different.

The wages at the ironworks are paid on a Friday. From Friday until Monday it constantly happens that the man is half incapacitated by drink. However good his resolutions, he is accosted on his way on pay-day by one tempting possibility after another. One of these possibilities, at any rate, seems to be very difficult to avoid. The wages, for convenience of handling, are paid as far as is possible in gold, an inconvenient form of currency with which to meet the daily expenses of the workman. If the pay-clerk has to give a sum of 30s. to 35s., he will pay the 30s. with a sovereign and a half-sovereign and the odd shillings only in silver. The workman will therefore have to change his gold coin or coins for silver ; and for this purpose the public-house, which takes care to be well provided with change at the end of the week, is the most obvious place. It is not surprising that some of the handful of change given to the workman by the publican should, partly in return for these kind offices, be expended on the

premises.   If the weekly wages could conceivably be
paid in silver instead of gold, so that the workman
had not a justifiable reason for entering the public-
house on his way home, there might be a good deal
less intemperance in the town.

In some of the lower quarters there are touts employed by the publicans, who are given drink for
nothing if they bring in a customer, and who stand
on the pavement and ply the workman with offers of
drink.   One man, who had to pass five public-houses
on his way home from his work, told a visitor that
he had to make a separate effort at each one ; he had
succeeded one winter's afternoon in passing four, and
if the fifth had not been there he 'would have got
home all right,' but at that last one he succumbed
and went in, cold, wet, and tired, to sit in the warm
and cheerful room, with the incidental subsequent
result that may be imagined.   If it had been
possible that this man should have found any other
equally bright and cheerful-looking place of resort
on his way, let alone five of them, he would have
had the chance, at any rate, of turning into one of
these instead, and would most probably have done
so.   But, unhappily, there are not many such places
other than public-houses, especially not many that
look attractive from outside.   The mere fact of that
blaze of light in an ill-lighted street is of incalculable
effect.   Some of the side streets in this town are so
dark at night that one cannot see the numbers on

the doors, and it is hardly possible to recognize the face of the passer-by. As the pedestrian turns down one of these from the more lighted thoroughfares, he becomes almost invisible by the time he has passed half a dozen houses. Many of the women who would hesitate to be seen the worse for drink by daylight slip out in the darkness to the nearest public-house, and stumble back again home. One evening at about eight o'clock, a visitor, groping down one of these streets, saw an inert mass on the pavement. It was so dark round her that she could not see clearly what it was until she knocked at a cottage door and asked for a light. She then saw it was the wife of one of the workmen, lying on the pavement helpless with drink.

The results may be less disastrous for the household when the children are sent to fetch beer than when the mother goes herself. The publicans in the town in question are forbidden to supply a child with beer or spirits in an open vessel; it must always be practically sealed or fastened. It is true that the child becomes familiarized with the public-house and its surroundings, but that familiarity is in most cases inevitable and only a question of degree.

All that has been said about the difficulty of combating drink applies equally to the habit of betting and gambling, which has taken an enormous proportion in the community we are describing, as it has, indeed, in every class of society all through the whole of the country. Among the ironworkers it is indulged in in various forms by men, women and children, with untiring zest.

The most prevalent form it takes is betting on horse-racing. Besides this, the men bet on billiard matches, on cards, on dominoes, on football matches. Betting on races is made fatally easy for them by the elaborate organization of the tipsters for supplying information and facilities, and by the extraordinary amount that is written about it in the papers. One page of the local paper is absolutely filled with the particulars of the race meetings; and it is not only in the local papers, but in influential daily papers, published in London, that the place left for Stop Press news of the latest important incident happening in the world is occupied by the name of the winner of a race, or by the odds on a horse.

The gambling propaganda is probably still more widely spread and more efficacious than that of drink. Bookmakers, forbidden to ply their trade in the open, and not allowed to stand in the public places as they used in the way of the man going to and from his work, have now been obliged to adopt other methods; they go from door to door and call at the houses, either when the men are at home or when they are out, and in the latter case they are quite as likely to gain an entrance as in the former.

The policeman may watch as much as he likes, it is difficult to stop every man who calls at a cottage door, or to prove that he is a bookmaker if he denies it. Added to which, most of the cottages have back doors and a way out into a back street, and nothing is easier than to go in at one door and out of the other. The bookmaker, therefore, is now playing a game of chance against the police, a new form of gambling added to the rest, in which, if he is successful, the enemy remains outwitted in the street. This view was confirmed a short time since by a constable who had been trying to run to earth some bookmakers suspected of having centres in two workmen's houses in a particular street. He said it meant that the man, besides wanting to bet for the sake of betting, wanted to outwit his adversary, and the policeman himself was conscious that he was embarked on a game. He said: 'I feel quite ashamed sometimes to think what I spend my time in doing and what I am after. I am neglecting my other duties. I am not thinking about the law or anything else. I'm just thinking, "I'll best that man"; and he is thinking he will best me.'

The fines inflicted on the man who is caught book-making are to the bookmaker a form of gambling loss to be accepted with other risks. This time he has lost, another time he will win; that is, he will get away without the police laying hands on him. And sometimes if he is caught the fine is paid by a 'gathering,' that is, a collection made among the men at the works.

I have no knowledge of what is the result in other countries of legitimizing and authorizing gambling by having public lotteries organized by the State, but it is conceivable that these should be a more wholesome outlet for the universal inevitable tendency than the surreptitious manner in which such operations are conducted in this country.

The systematic betting of the women, encouraged by bookmakers, is in many cases at first a quite deliberate effort, with or without the knowledge of the husband, to add to the income. A man comes to the door of a woman who, either from her own thriftlessness, or from stern necessity, is hard pressed for money, and presents her with the possibility of spending a shilling and winning £5. How should she not listen to him? She takes the bet, and it may actually happen that she wins; and if she does, small possibility that any subsequent arguments or exhortations against the practice should have any effect.

A well-meaning adviser, sick at heart at seeing the disasters brought about in the workmen's lives by betting and gambling, was earnestly entreating some of the women at the works to refrain from it, many of whom probably had listened to the bookmaker, one of whom had almost certainly made her house a centre for betting operations; but the admonisher was absolutely reduced to silence by one of the women saying, in a tone of heartfelt convic-

tion, ' But that £5 we won at the new year, it did
fetch us up wonderful.'   In face of the £5 suddenly
falling from the skies in the midst of a hard winter
into a household earning 30s. a week, it was difficult
to persuade the winners that the chances were that
it would not happen again.   The pathetic thing is
that the winnings, the sudden intoxicating winnings,
seem, and perhaps naturally enough, to go at the
best into useless channels, hardly ever to be laid
out in a way which would justify having won them ;
and yet part, no doubt, of the joy of embarking on
these perilous undertakings is the sheer insensate
joy of spending, and of spending on something that
has hitherto been as inaccessible as the Himalays.

On another occasion, after the visitors to a particular household had been perplexed for a long time
at the constant difficulties of a household where the
man was in receipt of 42s. weekly—difficulties which
were attributed to the thriftlessness of the wife—it
was one day revealed what was really happening by
the fact that when the visitor went into the kitchen
new possessions were being proudly displayed.   There
was a big shining toy drum hanging on the kitchen
wall, an object of pride and joy to the whole family,
the father and mother sharing in the almost awe-
struck exultation of the small possessor of the drum·
There was also a large new boiler to wash the clothes
in.   One could not find much fault with the way
the money had been expended ; one could find fault

only with the way it had been acquired. And yet, the most forcible argument against that way had been absolutely quashed by the result. Such a stroke of luck happening to one house in the street is bound to be an encouragement to others.

Another visitor had been in a house which was clean enough, but almost without the necessaries of life. She knew that the husband was steady and in receipt of good wages, that neither he nor his wife, a frail, anxious-looking creature, drank. The visitor could not imagine where the money went. While she was there one day two telegrams came in, and it then dawned upon her. She said to the woman : ' I see what it is ; you bet.' And the woman admitted it. She had begun it because she happened to be in desperate straits for money the first time a bookmaker came to her door; she had gone on with it since for the sake of the excitement and interest it brought into her life.

It was undesirable, no doubt, that this woman should bet—horribly undesirable, as the look of her house and surroundings showed. But what an illumination, albeit lurid, of outlook it must have brought to one, who before had had no personal interest in anything outside her sordid daily life, when she began to import into it these constant moments of alternating hope, fear, and wonder, when, day by day, as she took up the halfpenny evening paper to see the result of a race, she felt it might contain something governing her own destiny !

I am no metaphysician, and I am not competent
to discuss what is the underlying basis of the interest
inspired by guessing the result of a given uncertain
issue ; but it seems evident that such an interest,
visible in human beings from their earliest infancy,
does exist.

A child of a year old with whom its mother is
playing bo-peep will be twice as much amused and
excited if the mother does not always appear in the
same place, and lets it wonder where she will be
seen next. Half the games of childhood are based
on the excitement of uncertainty. We must all
remember the joy of throwing dice or spinning a
teetotum, in order to determine the advance of the
player in many childish games—race game, yacht
game, etc.—whatever they were called, variations of
the same, which consisted of moving something
round a board to get to a given goal. The joy, in
a more or less intensified form, of waiting for the
unknown to happen, seems to be radical in human
nature, and to be present in every age and at every
stage of existence ; and it is the more arresting when
we add weight and momentum to the excitement of
uncertainty, so to speak, by staking money on the
issue.

I would give a great deal to be furnished with
some brief, plain, convincing argument against
staking money on chance, which would seem to the
man of small means a cogent reason for giving it up.
To tell him that if he loses he will be ruined is not

final, since he will reply that if he wins he will be enriched. And every now and again a working-man does win, and thereby acquires in one moment a lump sum of capital that would be accessible to him in absolutely no other way. The fact that, as we are constantly told, the bookmaker in the long run is bound to win, matters little to the man who knows that, on one occasion at least, the bookmaker lost. Many of us cannot conscientiously express the belief that backing an opinion with money is a sin in itself, although we may long to suppress it on account of its results. We may feel the intimate conviction that the intense thrill of suspense purchased by it is pernicious and destroying to character ; that this form of redistribution of wealth adds not to the wholesome industry of the country, but to a traffic in unworthy excitement ; that on these grounds it is indefensible, and on the grounds of economics undesirable. We know that the individual is generally the worse for gambling ; that it tends to debase character, and to lessen the sense of responsibility. But these and other phrases of the kind are not very much good when addressed to men or women full of the excitement of the fray. Something terse, effective, powerful, is needed. The stock argument against a bet, that it makes one man's loss another's gain, may be said to apply in some degree to nearly all competitive forms of acquiring means and position, from the top of society to the bottom, from those who are following politics, art, or any of the professions, down to those engaged in the most petty details of commerce. Carrying off the prizes in any of these, perhaps means that the man who rejoices at having succeeded is likely to overlook the sorrow of the man who has failed. But this, let us hope, is not, and cannot be, so entirely destructive to character as the fierce, excited joy of the lucky gambler, whose success, purchased by no desert, no work of his own, may mean the absolute ruin in soul and body of the man who is sitting opposite to him.

The overwhelming argument against risking money on a bet, when the man who is betting has only just enough to live upon to begin with, is so patent that it hardly needs to be stated to him by anyone else ; for it is too plain that if he spends his money in buying the emotions of chance for himself he will not have enough left to buy bread for his wife and children. But if a man is not convinced by seeing his children hungry, he is not likely to be convinced by anything else ; he will reck little of character debased, standards blurred, and purpose ruined.

I cannot see what prospect there is of the tendency to betting being even modified, let alone stopped, except by trying to impress upon children, when they are still at the age which accepts a statement without requiring a reason for it, that it is undesirable to bet or to play for money. But this is not the point of view presented to the children in the community we are describing. Almost from baby-hood they play pitch-and-toss with pennies in the street; when they are sent across to the works with the men's dinners, they carry messages and communications from the tipsters and bookmakers as well; they take 'tips' to the workmen; they are given coins to bring back again and lay on a race; and all this is, from earliest childhood, a matter of course to them.

In one of the smaller telegraph-offices of the district, in which a visitor was standing, a small child came in who was just tall enough to reach the counter. Standing on tiptoe, it laid a telegram down under the visitor's very eyes. This telegram, written by the child's mother, was addressed to a tipster, and contained instructions to lay on a given horse.

I have no hesitation in saying that in this district, at any rate, a generation is growing up which is being deliberately trained in betting and gambling. The child who is given money to put on a horse for some one else very soon learns to do the same thing on his own account. It seems almost hopeless to try to stop a habit begun under conditions like these. Some of the schools in the town in question and other neighbouring manufacturing towns are in the habit of running excursion trains into the country on race-days, in order to get the children away from the racing surroundings altogether. Getting them away, if it could be done permanently, seems to be the only chance of dealing to any purpose with the question.

Card-playing in every spare moment is constantly
going on in the works, and quite young boys are
found furtively playing at every corner.   Playing for
money is supposed to be not allowed, and if
discovered it is fined.   The offence cannot be proved
unless one actually sees the money passing.   It may
be decreed at the works, or at a public-house or
private club, that the men may not play for money
or bet on the game, whatever it is, that they are
playing or watching; but no one can really prevent
them from doing so, as they can always, if necessary,
arrange the bet beforehand and settle about it afterwards.

'Handicaps,' especially in reference to billiard
matches, are very popular.   They are certainly a
sort of safety-valve where betting is concerned.   They
are organized thus : the men wishing to put into the
handicap contribute a sum agreed upon, usually a
few pence.   The whole amount of the contributions
is then spent in buying something to be played for—
it may be collars, ties, shirts, any article of clothing,
or, at Christmas time, a goose or some other form of
food.   This is gambling, no doubt, but it is, at any
rate, gambling to a fixed and limited extent only.

But betting on cards or on billiards, undesirable
as it may be, is nothing in comparison to the
disastrous amount which prevails of betting on racing.

I have the picture in my mind of a winter's day,
when the huge, cumbrous ferry-boat, which plies
across the Tees, taking the men backwards and

forwards from their work, was going slowly back in the late afternoon, laden to overflowing with the toilers who had finished their day's work. As it reached the shore some other men going across to the works from the town were waiting to board it, among them a man of about sixty, carrying his bag of tools in one hand and the evening paper in the other. He stood facing the crowd, as, with one question on their eager faces, they jostled across the gangway. And in one word he gave them the answer, the word they were waiting for—the name of a horse. It ran through the crowd like the flash of a torch, lighting up all the faces with a nervous excitement; and it seemed to the onlooker that there was not a man there whom that name did not vitally concern. That moment of tense expectation before the result was known was to many of the hearers part of the pleasure they were buying and paying for, some of them paying a price out of all proportion disastrous. One wondered how many of the homes to which those men were returning would be affected by the event of which they had just heard the news, how many of them, who might to-day be living in tolerable comfort and without anxiety, would be face to face with ruin.

There was a couple that, if not comfortably off, were succeeding in keeping abreast of their daily needs, albeit, with a struggle. The husband, return-ing home one Friday with his weekly wages in his pocket, met a bookmaker who gave him a ' tip,' said to be absolutely certain.   The workman at once believed him and staked the whole of his wages upon it.   He came home in a great state of excitement.   His wife was standing on the threshold.   He told her what he had done.   The wife received the news with the most eager approbation, said what a pity it was he had not had more to put upon it, then had an inspiration, and suggested pawning all the furniture. They then, having pawned the furniture, staked every penny of the proceeds on the horse, and subsequently saw in the paper that they had lost. On the Monday morning, when the moment of settling came, they were absolutely destitute.

We deceive ourselves if we assume that the possible disastrous results of gambling are always enough to deter people from doing it.   In some cases, no doubt, where the habit is not firmly established, and where betting has been embarked upon with the deliberate purpose of making money, ill-luck may act as a deterrent.

The wife of Peter S., a man who received 37s. a week, told a visitor that he had always betted until he lost a large sum at Stockton Races, after which he never betted again.   Another man, Charles R., said that before he had children he used to bet, but when he had 'several little uns' he made up his mind he couldn't afford it.   It is well when losing has this effect.   But to the confirmed gambler, the result of the luck being against him is simply a frenzied determination to go on until it turns.

And the lookers-on try in vain to stem the tide, see one tragedy after another that no power, no persuasion seems able to prevent, and feel wellnigh hopeless as to the possibility of arresting this great disintegrating process. Many ways have been tried by the well-meaning, but the forces arrayed against them are overwhelming. It is a terribly unequal fight that is being waged. The crusade against it, ardent though it may be, encounters as ardent a propaganda in favour of it, in which a far greater number of people are engaged than on the other side. The combatants on the side of gambling include all the immense number of professionals, who are deliberately spreading gambling and betting, in one form or another, through the land for the sake of gain and to earn their livelihood; all the amateurs, who may not do it for the same reason, but who devote their time and thoughts to it with a concentration which make them almost professionals; all the great mass of the community from top to bottom, including the hundreds of thousands of the working classes, who, without specializing in it to this extent, do gamble and bet whenever they have the opportunity, and play for money nearly every day of their lives. All these are practically incessantly engaged in furthering gambling. Then, besides, there are the many, many seemly and well-conducted households, the members of which do not make a practice of playing for money, but enjoy doing so if it comes in their way. And there are a great many people who, not caring about cards at all and having no tendency to gambling or betting, are still quite indifferent to the fact that other people should indulge in it, even to excess. The net result is that among the people who do not care about it and do not indulge in it, only a certain number feel strongly enough on the question of gambling to endeavour actively to stop it. They are therefore in a hopeless minority. I fear that unless the world changes very much there will always be an overwhelmingly greater number of people actuated by the love of gain and of pleasure than of people who, striving disinterestedly to fulfil high ideals of conduct, are endeavouring to lead others to do the same. Their task is wellnigh superhuman. The wretched truth is that those who are trying to stop gambling and betting are mostly in the position of preaching to the converted. The unconverted simply do not listen. It is the teaching of observation and experience that in any class the people who actually refrain from gambling themselves, although they may see the necessity for legislating against it, are usually, although not invariably, those who have no tendency to it.

Most people of the well-to-do classes do not make a crusade about anything; certainly not about this subject, and for obvious reasons. It would be super-fluous to insist further upon the extent to which every class of society is permeated with gambling; but in the case of the people who can more or less afford to lose—to whom losing, at any rate, does not bring any conspicuous misfortune, privation, or degradation—it is more difficult to give a convincing reason why they should leave it off. In these days especially, when nearly every one in the middle classes as well as fashionable society plays Bridge, playing cards for very small sums of money, as is often the case, seems an innocent amusement enough, and not more unprofitable than any other of the idling occupations of leisure which would be likely to present themselves as alternatives to it. A widely-read halfpenny paper has recently been publishing a correspondence on whether Bridge 'pays.' Various people have given their experiences, and have shown on the whole that it does pay. Whether this array of testimony is authentic or not, it certainly gives a general view of the practice of to-day. One can imagine that to read the columns on this subject, written presumably by the educated, to look at the later page of the paper, quite full of the lists of race-meetings, must go far in the workman's mind, and in many another, to neutralize the effect of the less numerous and less conspicuous utterances against betting and gambling put forth either in print or in speech.

Many of us feel that those who, if they bet or play for money, can afford to lose, would do well all the same to refrain from playing for the sake of example; for the only chance of lessening the evil seems to be that a gradually spreading and increasing pressure of opinion should grow up against it in all layers of society.   But that pressure of opinion seems so far to exist least of all amongst those whose example would be most conspicuous and most powerful.   It appears to be too much to hope that any of these, to whom playing for money is the everyday resource of their leisure, should relinquish their favourite amusement on the chance of lessening temptation to some unknown workman out of their vision.   To get as far as this, even, is commonly not within measurable distance, for there are many people whose lot in life, to their fortune, or misfortune, has not brought them into contact with the object-lessons given by the life of the working classes, and who are serenely unconscious of the perils that lie along these less fortunate paths.   When a Select Committee was appointed some years ago to inquire into the increase of public betting among all classes, and into the possibility of checking it, one of the most highly placed and influential persons in the country gave his opinion as follows :

He thought there was nothing wrong or immoral in betting.   He would very much regret its being stopped ; it would certainly injure the national amusement of horse-racing.   He thought betting the support of racing.   He saw nothing wrong in the bookmaker's profession, and, in reply to a question as to their taking small sums from children in poor neighbourhoods, he said he had no knowledge of that sort of betting.

This evidence was in startling contrast to that of other witnesses who did know what was going on, and what is going on every day among the working classes ; and it is not encouraging reading to those of us who believe that it is the class of society to which this witness belongs from which, if at all, the mighty effort, the concerted action to suppress gambling, ought to come. We may well feel hopeless as to the chances of 'suppressing, or even modifying, that which the great majority of the nation— those alike whose children starve if their parents gamble, and those whose children never know a privation—are upholding and fostering with all their might.

Do not let us be under any illusion, those of us who look on at a great social evil we dream we may reform. The evils discussed in this chapter, that bring in their train not physical deterioration only, but spiritual deterioration as well, are not the monopoly of one particular class, and the tendency to them may be latent in us all. Each of us is born into the world as a mass of possibilities, and it is the lot in life and the surroundings of the average mortal that determine which of these possibilities, whether for good or for evil, shall be developed, and which are to remain for ever dormant.

A fierce light beats in these days upon the working classes, revealing much that in more prosperous quarters is not seen ; but it is probably there all the same. From the fact that so far, unfortunately, it has not been the custom to investigate, tabulate, report upon the private and individual lives of the well-to-do, it has come to pass that the working class is used, so to speak, as the unit of the moral investigation, until we wellnigh believe that that class is the chief repository of the vices and virtues of the nation. For these are bound to 'crop out,' in geological parlance, to the day, without being overlaid with concealing matter; they are immediately recognizable by their tangible results, and can at once be reduced to the damning and unanswerable terms of pounds, shillings and pence—so definitely indeed, that we are in the habit of labelling as 'good,' according to a somewhat elementary moral standard, those among the working class who do not need, or do not ask, to have their means supplemented from outside.

A great French writer has said that the cardinal difference between the lot of the rich and the poor is that the former have more margin in which to remedy mistake and misfortune. In studying the lives of the workers described in this book, we become convinced of the truth of this saying. The path the ironworker daily treads at the edge of the sandy platform, that narrow path that lies between running streams of fire on the one hand and a sheer drop on the other, is but an emblem of the Road of Life along which he must walk. If he should stumble, either actually or metaphorically, as he goes, he has but a small margin in which to recover himself.

But it is good to think that even in the face of these conditions—it may be partly, indeed, because of them—many of the ironworkers of the North hold strongly and undauntedly on their way; that this relentless environment often succeeds in fashioning an admirable human product, and that among these ironworkers, as all will know who have frequented them, are to be found some of the finest and most interesting types of working-men in the kingdom.

It is probable that any human being attempting to describe the life of another will only approximate to representing that life as it appears to the person described. A good deal of guessing will always remain to be done, and at the end we may not know whether we have guessed aright—whether we have understood or misunderstood. But it is always worth while to try to understand; we shall learn more, at any rate, in that way, than if were are content only to guess.

If any of the workmen whose lives are here described should come to read this book, let them judge it with kindly tolerance, and with that friendship—heartily reciprocated—which they have for so many years extended to the writer.

THE END